# Farmer Buckley's Exploding Trousers

And other odd events on the way to scientific discovery

**NewScientist**

# Farmer Buckley's Exploding Trousers

## And other odd events on the way to scientific discovery

Edited by
## Stephanie Pain

PROFILE BOOKS

First published in Great Britain in 2011 by
Profile Books Ltd
3A Exmouth House
Pine Street
Exmouth Market
London EC1R 0JH
*www.profilebooks.com*

10 9 8 7 6 5 4 3 2 1

A CIP catalogue record for this book is available from the British Library.

ISBN 978 1 84668 508 8
eISBN 978 1 84765 749 7

Text design by Sue Lamble
Typeset in Palatino by MacGuru Ltd
*info@macguru.org.uk*

Printed and bound in Great Britain by
Clays, Bungay, Suffolk

The paper this book is printed on is certified by the © 1996 Forest Stewardship Council A.C. (FSC). It is ancient-forest friendly. The printer holds FSC chain of custody SGS-COC-2061

**FSC**

**Mixed Sources**
Product group from well-managed
forests and other controlled sources

Cert no. SGS-COC-2061
www.fsc.org
© 1996 Forest Stewardship Council

# Contents

# Introduction

On 12 August 1931, New Zealand farmer Richard Buckley hit the headlines of his local newspaper. What had he done? Strictly speaking, he hadn't done anything. He owed his claim to fame to his trousers. One minute they were hanging harmlessly before the fire to dry. The next there was a huge blast and a ball of flames. Farmer Buckley's trousers had exploded. Buckley was an early victim of what soon became an epidemic of exploding trousers. All across the country, there were reports of detonating dungarees and blazing breeches. Buckley had been lucky. He hadn't been wearing his at the time.

If you haven't heard of Farmer Buckley's near-lethal legwear, you're in good company. The incident quickly faded into obscurity, and as far as we know it has never featured in any book on the history of science. Here at *New Scientist*, though, we think those trousers deserve a good airing. Why? Because the story is a salutary one, an example of how scientific advances meant to improve people's lives and livelihoods can sometimes backfire – in this case literally. In its own small way, the tale helps to counter the widespread misconception that science and technology have advanced along a straight and narrow road from one milestone to the next.

They haven't. The quest to understand and manipulate the world we live in has been anything but straightforward. Our scientific ideas and the technologies we rely on today have been arrived at by a process best described as haphazard.

One brilliant idea doesn't necessarily follow another, nor does one breakthrough lead inexorably to the next. The search for understanding meanders here and there, stumbles up blind alleys and is frequently never seen again. Nor, disappointing though it might be, are all those who endeavour to push science and technology beyond their known limits simply in it for the kudos, happy to satisfy their curiosity and maybe earn a place in history. There are baser urges at work. Some are driven by greed or commercial expediency. Some are gripped by an obsession, while others merely indulge their whims and foibles.

Newton might well have believed that he saw further than others by standing on the shoulders of giants, but giants are rare. Most of what we know has been built slowly and ponderously on the efforts of more diminutive predecessors. Of course there have always been plenty of smart, resolute, even brave scientists with imaginative and ingenious ideas. But the history of science is also peppered with a cast of fascinating characters including money-grubbers, fraudsters and, dare we say, cranks whose ideas went nowhere, held back progress or even did harm. And so, to set the record straight, we'd like to introduce you to some unfamiliar characters and events that helped make history.

As much as we admire the Newtons and Darwins of this world, we think it's unfair that quite so much attention is lavished on the geniuses and their triumphs. They've had centuries of fame. Instead this book, which started out as a series of snapshots of people and events that first appeared in *New Scientist*'s popular Histories column, is devoted to those that are all but forgotten – but shouldn't be. That's why you won't find any mention of Newton's apple or Darwin's finches – but you *will* learn what happened to Farmer Buckley's trousers.

Stephanie Pain

# 1 No pain, no gain

Where would we be without all those pioneers who put their bodies on the line in the name of science? Not all are as self-less as they seem. Sometimes one body feels the pain and another makes the gain. Just occasionally both pioneer and patient get something out of it.

## ✺ Scalpel, suture, salt beef

*Alexis St Martin was a dead man: of that everyone was certain. When the young Canadian trapper entered the general store of the American Fur Company in the northern frontier town of Mackinac on the morning of 6 June 1822, he had no booze-fuelled fights to pick, no slights to avenge, no pelt deals to dispute. He was simply in the wrong place at the wrong time – directly in front of a trapper's shotgun when it accidentally went off. 'The whole charge entered his body,' a witness reported. 'The muzzle was not over three feet from him – I think not over two. The wadding entered, as well as pieces of his clothing; his shirt took fire; he fell, as we supposed, a dead man.' But the shooting of Alexis St Martin was to take a bizarre turn, one that would help to unravel the mysteries of digestion.*

There was a terrific blast, a groan and then a thud as Alexis St Martin's body slid to the floor. No one who witnessed the accident really believed a doctor could help, but they sent for one anyway. William Beaumont, an army surgeon based at the local fort, arrived some 20 minutes later to examine a dreadful wound under the victim's left breast. The shot had blown off 'integuments and muscles the size of a man's hand... materials forced from the musket, together with fragments of clothing and pieces of fractured ribs, were driven into the muscles and cavity of the chest'. Beaumont cleaned the gaping wound as best he could, but with the trapper's breakfast running out from a hole in his stomach, the end seemed certain. 'The man can't live 36 hours,' he remarked as he left.

Beaumont's prognosis was based on hard-won experience. As assistant army surgeon during the war of 1812, he had treated soldiers with appalling injuries. After one engagement he spent two days amputating limbs and trepanning broken skulls. By the time he examined St Martin, Beaumont had seen all imaginable wounds, and their typically fatal results.

Yet the stubborn trapper survived the loss of blood, 10 days of fever, and weeks during which shattered bone and cartilage oozed out of his wound. Eating and drinking were impossible; the contents of his stomach simply spilled out from the hole. For weeks, the only way to sustain St Martin was by injecting fluids into his rectum. But he rallied, clinging to life for so long that Beaumont was forced to take the penniless patient into his own home. A few weeks later, the hole had stabilised sufficiently for the surgeon to cover it with bandages, allowing the trapper to eat.

Three years passed under Beaumont's care, and St Martin could live and work much as before. There was just one problem: although the wound had healed, there was still a hole, or fistula, leading straight into his stomach. With the bandages off, St Martin's gastric fistula allowed Beaumont to view inside the stomach 'to a depth of five or six inches' and watch food and drink entering it. The gunshot wound had healed into a window on the workings of the stomach – and offered Beaumont an extraordinary scientific opportunity.

Digestion was still poorly understood in the 1820s. For centuries, anatomists assumed food was broken down by a process of fermentation or 'putrefaction' by some hazily defined 'vital spirit' in the stomach. By the 1790s, the idea of chemical breakdown was becoming better accepted. In one experiment, French scientist René Réaumur fed sponges to his pet bird; after the bird obligingly vomited these back up, Réaumur squeezed the gastric fluids out onto litmus paper, showing it was acidic. Another researcher, French physician Antoine Montègre, had an extraordinary patient who could apparently vomit at will, and he declared in 1812 that the resulting liquid was not acid, but nothing more than swallowed saliva.

In 1825, Beaumont moved to a new post in Niagara, New York. He took St Martin with him and began a groundbreaking series of experiments. He tied food – including a piece of bread, some cabbage, and a lump of salted beef – onto a silk string and lowered it down the hole into St Martin's stomach. He left the morsels for carefully measured intervals before reeling them back up to examine the results. Beaumont also inserted a thermometer to determine the stomach's temperature, which was 100 degrees Fahrenheit (38 °C), and used a tube to draw out 'gastric juice' for further study. He quickly discovered two key facts: first, gastric juices were acidic, and second, they were not present when the stomach was empty. Food stimulated their production, although they also sloshed about when St Martin got angry.

Beaumont found that while certain foods, such as boiled rice, broke down after an hour in the stomach, salted pork could take 4 or 5 hours. Neither broke down in tubes of gastric juice outside the body unless they were warmed up. Larger pieces took longer to digest than smaller ones, and a little shaking – just as he'd seen the stomach do with its squirming – also worked wonders. So food was broken down by rendering, heating and shaking: it was, in other words, a typical reaction with a solvent. There was no need to invoke some vague 'vital spirit'.

In July 1825 Beaumont received two months' army leave, allowing doctor and patient to visit medical friends in New York and Vermont. But for St Martin, a hard-drinking, illiterate labourer, gastroenterology held no charms. He wearied of people gawping at his guts. The nearby Canadian border beckoned and one day he slipped away.

Nothing more was heard of him until three years later, when Beaumont received a letter from an agent of the American Fur Company: 'While in Canada last winter I succeeded in finding your ungrateful boy, Alexis St Martin. He is married... poor and miserable beyond description, and his wound is worse than when he left you.' Beaumont was thrilled and paid for St Martin and his family to come to live with him at his new posting in the Michigan Territory. St Martin was to serve as Beaumont's test subject for the next four years.

Beaumont threw himself into his work. He sent out bottles of gastric fluid for analysis, determining that a key ingredient was hydrochloric acid. He carried out hundreds of trials dangling food into St Martin's stomach, carefully examining the texture and weight of the semi-digested foods he fished out. His scientific fortitude knew no bounds: he even examined the stages of digestion by sampling St Martin's gastric fluids in his own mouth.

The result of these labours was Beaumont's *Experiments and Observations on the Gastric Juice and The Physiology of Digestion*, published in 1833. The first great work of physiology from the USA, it established the modern understanding of digestion.

St Martin eventually returned to Canada, where he was courted by doctors as a test subject, to no avail. Despite alcoholism, poverty and a hole in his stomach, St Martin outlived his doctor by 27 years, and outlasted 12 of his own 17 children as well. When he died near Montreal in 1880, his family kept his body rotting in the June heat for four days, so that nosy doctors would not attempt one last examination. As an extra precaution, they then buried a very ripe Alexis St Martin in an unusually deep grave. And at long last, his stomach stayed covered up for good.

## ☀This won't hurt a bit

*If you want to make it into the history books as a hero of medical science, you can't beat a bit of experimentation – on yourself, that is. Is a new drug safe? Take some and find out. Does that vaccine work? Try it and see. The only catch is that you have to survive the experiment long enough to write up your results in a suitably eminent medical journal. One man who did, and earned worldwide fame, was the German surgeon August Bier. In 1898, Bier invented spinal anaesthesia. After a few promising tests on patients, Bier wanted to find out how much they felt during an operation and why they developed horrible headaches afterwards. So, one summer's evening, he asked his assistant to anaesthetise him. It was an experiment they might have preferred forgotten.*

The two surgeons had finished work for the day. But instead of going home, they began to prepare for one more operation – a little out-of-hours experiment intended to advance the art of anaesthesia. August Bier was a rising star at the Royal Surgical Clinic in Kiel. His young assistant, August Hildebrandt, had agreed to help him.

What happened next was not so much heroic as comic. Just one little mistake and courageous selflessness turned to black comedy. It made Bier's name. But the events of that evening would be forever etched on Hildebrandt's memory, not to mention several other parts of his body.

In the 1890s, general anaesthesia was decidedly dodgy. Chloroform sent patients gently to sleep – but there was no room for mistakes. A few drops too many and the patient would be dead before the surgeon picked up his scalpel.

Ether wasn't quite so dangerous, but it was slow to act – surgeons sometimes started to operate before their patients had gone under. The survivors suffered unpleasant side effects – from violent headaches and vomiting to ether pneumonia.

Bier reasoned it should be possible to banish sensation from most of the body without knocking the patient out completely by injecting a small dose of cocaine into the cerebrospinal fluid that bathes the spinal cord. He tried his technique on half a dozen patients. They lost sensation from the lower part of their bodies long enough for him to carve out chunks of diseased bone from their ankles, knees and shins and even the thigh and pelvis. 'On the other hand, so many complaints had arisen in association with this method that they equalled the complaints usually occurring after general anaesthesia,' he wrote. 'To arrive at a valid opinion, I decided to conduct an experiment on my own body.'

The procedure was simple enough. Hildebrandt had to make a lumbar puncture by plunging a large needle through the membranes that protect the spinal cord into the fluid-filled space beneath. Then he had to fit a syringe on the needle and inject a solution of cocaine. But preparations for the experiment had been less than meticulous.

Hildebrandt made the lumbar puncture. Then, with his finger over the hub of the needle to prevent fluid from leaking out, he took up the syringe of cocaine – only to find it was the wrong fit. As he fumbled with the needles, Bier's cerebrospinal fluid began to squirt out. Horrified, Hildebrandt stopped and plugged the wound. This was when the pair should have called it a day. Instead, Hildebrandt offered to take Bier's place.

At 7.38 pm, after checking the needles more carefully, Bier began. The cocaine worked fast. 'After 7 minutes: Needle pricks in the thigh were felt as pressure; tickling of the soles of the feet was hardly felt,' he noted. Bier jabbed Hildebrandt in the thigh with a needle. Nothing. He tried harder, stabbing the thigh with the surgical equivalent of a stiletto. Still no response. Then, 13 minutes into the experiment, Bier stubbed out a cigar on Hildebrandt's leg.

Bier now wanted to know how far the insensitivity extended, and invented a simple test. 'Pulling out pubic hairs was felt in the form of elevation of a skinfold; pulling of chest hair above the nipples caused vivid pain.' So now he knew. It was more than 20 minutes since Hildebrandt had stopped feeling pain. How much more could he take? Bier increased his efforts. He smashed a heavy iron hammer into Hildebrandt's shin bone and then, when that failed to have any effect, gave his testicles a sharp tug. In a final burst of enthusiasm, Bier stabbed the thigh right to the bone, squashed hard on a testicle and, for good measure, rained blows on Hildebrandt's shin with his knuckles.

After 45 minutes, the effect of the cocaine began to wear off. The two surgeons, one missing a significant amount of cerebrospinal fluid, the other battered, burnt and suffering from serious stab wounds, went out for dinner. 'We drank wine and smoked several cigars,' wrote Bier.

The next morning, Bier woke feeling bright and breezy. By the afternoon he had turned pale, his pulse was weak and he felt dizzy whenever he stood up. 'All these symptoms disappeared as soon as I lay down horizontally, but they returned when I arose. In the late afternoon, therefore, I had to go to bed.' He stayed there for the next nine days. When he finally got up again he felt quite well. 'I was perfectly able to tolerate the strain of a week's hunting in the mountains,' he wrote.

Hildebrandt didn't escape so lightly. The first night he was violently ill. He had a splitting headache and was sick. But someone had to tend to the clinic's patients and, with Bier in bed, the job fell to him. Each morning for the next week, Hildebrandt dragged himself to work. Each afternoon, he staggered home and collapsed into bed. 'Dr Hildebrandt's legs were painful, and bruises appeared in several places,' wrote Bier, rather understating the case.

When Bier wrote his ground-breaking paper describing the experiment, he gave a blow-by-blow account of what Hildebrandt had endured. As far as Bier was concerned, the experiment was a huge success. He had shown that a tiny dose of cocaine could deaden sensation for long enough to perform a major operation. Spinal anaesthesia was far safer than general anaesthesia, and within two years surgeons around the world were using it. Bier put the headaches down to the loss of cerebrospinal fluid, and he was right – this was finally proved in the 1950s.

Hildebrandt, though, had gone right off Bier and became one of his most vehement critics. When a row blew up over who had really been first to invent spinal anaesthesia, Hildebrandt championed Bier's rival, an American neurologist called James Corning. Hildebrandt never said why. Perhaps he was shocked by the zeal with which Bier had battered him. Maybe he was miffed because in the end Bier was recognised as a pioneering surgeon, while he was forever known as the man whose boss had tugged his testicles.

## ☀ The accidental aeronaut

*'I have been insensible,' said the scientist. 'You have,' said the aeronaut, 'and I too, very nearly.' Meteorologist James Glaisher and balloonist Henry Coxwell had just survived a trip into the stratosphere. Squashed into a wicker basket dangling from a huge gas-filled balloon, they had inadvertently risen to an altitude of 11,278 metres – higher than anyone had ever travelled before. Glaisher lost consciousness, while Coxwell, on the verge of passing out and with badly frostbitten hands, desperately tugged the valve line with his teeth to let out gas and bring the balloon to a less lethal altitude. As Coxwell later admitted, he had been more than a little anxious about his distinguished passenger: 'I began to fear that he would never take any more readings.' Yet, within minutes, Glaisher picked up his pencil and resumed his observations.*

When Henry Coxwell, one of Victorian England's most famous balloonists, agreed to take a scientist high above the Earth, he hadn't banked on taking one of the country's most distinguished meteorologists. It was 1862, and James Glaisher had been superintendent of the Magnetical and Meteorological Department at Greenwich Royal Observatory for more than 20 years. He was middle-aged, meticulous and methodical, a man who made measurements, organised observers and sat on important committees – scarcely the sort suited to aeronautical adventures, Coxwell suspected. Within months, however, the two men would be national heroes, with the public agog for news of their every ascent.

As Coxwell soon discovered, Glaisher was not after adventure. To him, a balloon was simply a means of extending his observations into the upper reaches of the atmosphere. Nor had Glaisher planned to be the one in the balloon. As a member of the balloon committee of the British Association for the Advancement of Science (BA), he had been pushing for a series of high-altitude ascents for years, but no one had a big enough balloon.

Then, in 1861, Coxwell offered to build a suitable balloon if the BA agreed to hire it for its scientific flights at £50 a time. The BA jumped at the offer and so Coxwell built a monster balloon which he named 'Mammoth'. Finding the right observer to fly in it proved harder, though. No one met Glaisher's high standards – so he volunteered himself.

On 17 July 1862, Glaisher discovered what he had let himself in for. The 'Mammoth' was to take off from the gasworks at Wolverhampton in the English Midlands. The windy weather at first made it difficult to inflate the balloon, and then a gust whisked it up before Glaisher had fixed his delicate instruments into position. This, he wrote, 'was by no means cheering to a novice who had never before put his foot in the car of a balloon'.

A few minutes later, 'Mammoth' emerged above the clouds into brilliant sunshine and Glaisher managed to settle down to work. Coxwell's job was to take the balloon as high as possible. Glaisher's was to record the temperature, air pressure and moisture, make notes on cloud formations and measure the speed and direction of any air currents they encountered – and to check the reliability of the new-fangled aneroid barometer.

By the time they reached 6,000 metres, the men's lips had turned blue and they could hear the pounding of each other's hearts. At 7,500 m they found breathing difficult. At such an altitude, Glaisher wrote, 'it requires the exercise of a strong will to make and record observations'. By 9,000 m, he had made an important discovery. Measurements made on mountains had suggested temperature dropped uniformly with altitude, around 0.5 °C for every 100 m. On this ascent, though, the drop was anything but uniform: first the temperature plummeted, then it fell more slowly, then quite unexpectedly rose before dropping sharply again.

By now Coxwell feared the balloon was about to head out to sea and he brought it down, landing with a crash that broke Glaisher's instruments. The meteorologist, however, had proved as heroic as any aeronaut.

Over the next three years, Glaisher made 27 flights. Not all were high-altitude ascents organised by the BA. Coxwell's balloons were a summer fixture at the Crystal Palace pleasure grounds in the London suburb of Sydenham. Among the millions who flocked to see the great glass palace, there were plenty willing to pay for a balloon ride over the city. The balloon ground had its own supply of coal gas, piped up the hill from Sydenham gasworks. The gas company's engineer even tweaked the mix to provide a lighter gas for lifting balloons. This arrangement was so convenient that Glaisher did make some official high-altitude ascents from Crystal Palace, but in his hunger for data he sometimes hitched lifts on Coxwell's pleasure trips, squeezing himself and his instruments into the basket with up to a dozen others.

Summer daytime flights left big gaps in his data, so

Glaisher also risked ascents at night and in winter. Whatever the time, every flight was dangerous: he survived several crash landings, had a close shave with a cathedral spire, and more than once faced the prospect of ending up in the sea. But it was one particular flight – on 5 September 1862 – that turned Glaisher and Coxwell into celebrities.

They took off from Wolverhampton gasworks just after 1 o'clock. The balloon rose through a thick bank of cloud then emerged into bright sunlight 'with a beautiful blue sky, without a cloud above us, and a magnificent sea of cloud below, its surface being varied with endless hills… and many snow-white masses rising from it'. The balloon was rising extremely fast, and spinning as it went. They reached an altitude of 8,000 m in 47 minutes. 'Up to this time I had taken observations with comfort,' recalled Glaisher.

Then, as the balloon rose higher, he found it hard to read his thermometer and asked Coxwell for help – only to find him gone. During their giddy ascent, the valve line that allowed Coxwell to control the gas in the balloon had become tangled, so he had climbed up into the hoop above the basket to free it. They continued to rise. At 11,000 m, Glaisher realised he couldn't move his arms or legs. His head fell sideways. 'I dimly saw Mr Coxwell in the ring and endeavoured to speak, but could not, when in an instant an intense darkness came. I thought that I should experience no more, as death would come unless we speedily descended.' By this time Coxwell's hands had frozen to the metal hoop. He grabbed the valve line in his teeth, and by nodding his head released some gas. The balloon began to descend.

'Do try – now do!' Glaisher realised someone was speaking to him. He opened his eyes and sat up. 'I have been insensible,' he said. Coxwell agreed he had. 'I recovered quickly,' Glaisher reported in *The Times*. 'But Coxwell said, "I have lost the use of my hands; give me some brandy to bathe them."' Glaisher poured brandy on Coxwell's blackened hands, then picked up his pencil and resumed his observations. They had, Glaisher reckoned, reached the limit of human existence.

In 1869, Glaisher made a further 28 ascents, this time in Henri Giffard's great Captive Balloon, which was tethered in a London park. During his earlier ascents, Glaisher had rarely had time to take readings at lower altitudes because the balloon rose too quickly, but Giffard's balloon was controlled by a steam-powered winch, which meant he could stop to take readings. Glaisher might have made more flights, but later that year the Captive Balloon escaped and crashed.

Glaisher's balloon exploits proved fruitful, however. He refuted the idea that temperature dropped steadily with altitude, and discovered that the atmosphere was like the sea, with currents moving at different speeds and in different directions, some warm and some cold. He found that conditions varied with time of day and season, and proved aneroid barometers were reliable. He also learned from experience that England was probably the worst place in the world to do science in a balloon. No matter where you started from, all too soon you were at risk of floating out to sea.

## 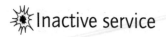Inactive service

*The sea is grey and choppy. There's a splash. A body floats into view. An airman, forced to bale out of his burning plane, is bobbing up and down in the waves. His life jacket keeps him afloat, but he's lying face down, arms and legs dangling helplessly beneath the water. In the standard wartime movie, this is the point at which the audience finally realises that the hero isn't coming home, the sun sinks into the sea and the credits roll. But this film is a record of a scientific experiment. Instead of the setting sun, a young man in a one-piece swimsuit jumps into the water and pulls out the drowning man. With the help of a wave machine, the swimming pool at Ealing film studios in west London does a good impression of the*

*North Sea on a squally day. Warships have sunk in this pool. Spit-*
*fires have crashed into it. And actors playing heroes have slipped*
*quietly beneath its waves. Cleared of model ships, planes and actors*
*with stiff upper lips, the pool is an ideal place to perform tests that*
*will contribute far more to the war effort than a few morale-boosting*
*propaganda films.*

The real Mae West never failed to turn men's heads. Unfortunately, the Mae West life jacket couldn't be relied on to do the
same. During the Second World War, airmen who baled out
over the sea were often unconscious when they hit the water.
And by the time rescuers reached them they were often dead.
Their life jackets kept them afloat – but if they landed face
down they drowned.

Edgar 'Gar' Pask was shocked by how often this
happened. Pask was a young anaesthetist at the University of
Oxford. But in 1941 he left his job to join the Royal Air Force.
The RAF immediately sent him to work at its physiological
research centre at Farnborough. During a stint as an observer
on a rescue launch, he became increasingly disturbed by the
pointless deaths. If an unconscious man was to have any
chance of survival, then his life jacket must twist him around
so that he floated face up with his head held clear of the water.

Ever since he'd joined the RAF, Pask had been working on
ways to increase airmen's chances when things went wrong.
There were plenty of suggestions for ways to modify the Mae
West. But they needed testing. That ought to be easy enough:
send a man into the deep end, tell him to relax and pretend to
be unconscious and let the life jacket do its stuff.

Only it wasn't that simple, as Pask discovered when he
tried it himself. 'It has been found impossible for the conscious
volunteer to remain quite inert, particularly when things
seem to be going wrong,' he admitted later. He insisted that
the tests must be done on an unconscious man – and that he
would be the guinea pig.

Pask, dressed in full pilot's gear of battledress, service

underwear and a thick woollen sweater, was anaesthetised deeply. He was connected to an anaesthetic machine at the poolside by a long, buoyant hose attached to a tube pushed down his windpipe. The tube was held tightly against the wall of his windpipe by an inflatable cuff. Robert Macintosh, Pask's boss from Oxford, controlled the flow of ether while a team of swimsuited scientists repeatedly threw the unconscious man into the pool.

The trials seemed endless. Some were in fresh water. In others, sacks of salt were thrown in to increase buoyancy. And finally, the team decamped to the film studio at Ealing to see how effective the redesigned life jackets were in rough water. In the movies, model ships appeared to battle against immense seas. In reality, the waves generated by the studio's wave machine were only a metre high. But that was enough to swamp a half-submerged man in a badly designed life jacket. There was no doubt that Pask was in danger. There was always the risk that water might seep past the cuff around the tube in his windpipe before the observers realised he was in trouble.

By now, however, Pask was used to the rigours of research with the RAF. His first mission had been to identify a survival suit that could cope with the freezing waters of the far north Atlantic. Such a suit was vital for any pilot assigned to protect the northern convoys. If they were shot down, they would last only a few minutes in those waters. Pask tested the designs himself, jumping into the sea off the Shetland Islands where the water was freezing and strong winds increased the chill factor. The tests were evidently a success: while Pask complained that he was too warm, his team of observers had to call a halt to the tests before they froze to death.

In 1942, Pask put himself at even greater risk. The team at Farnborough had been asked to answer something all aircrew desperately wanted to know. What was the maximum altitude they could bale out at without breathing apparatus and survive? The higher you went, the lower the air pressure.

Too high and there wasn't enough pressure to drive oxygen through the air sacs of the lungs and into the blood.

For these tests, Pask dangled from a scaffold in a parachute harness and breathed mixtures of gases with successively lower percentages of oxygen to simulate pressures at different altitudes. At the highest 'altitudes' – between 35,000 and 40,000 feet – he lost consciousness, his muscles twitched and he struggled to breathe. It was life-threatening stuff, but the experiments established that pilots could parachute from around 35,000 feet without oxygen apparatus and still have a reasonable chance of survival.

Most dangerous of all Pask's escapades was a series of experiments in 1943 to find out the best method of resuscitating a drowned airman in a speeding rescue launch. Some people recommended the mouth-to-mouth method. The Royal Life Saving Society preferred the Schafer method, laying the victim face down and applying pressure to the lower back. Others used the Sylvester method, lifting the arms and pushing down on the chest. Finally there was Eve's technique, in which rescuers tilted or rocked the victim back and forth. But which was best? Which forced the most air through a drowned man's lungs?

For Pask, the only reliable way to find out was to compare all the methods on the same person under the same conditions. And that person must have stopped breathing. Again, Pask offered to be the guinea pig. This time, he would be anaesthetised to the point of respiratory arrest. Unless his breathing was restored he would die.

During two days of tests, Macintosh anaesthetised Pask until he stopped breathing, then connected a tube in Pask's windpipe to a device that would monitor the amount of air going in and out of his lungs during the attempts at artificial respiration. On the first day, in the course of two hours of experiments, Pask was taken to the point of respiratory arrest seven or eight times. Two days later he went through it all again. The trials showed that Eve's rocking method pumped the most air through the lungs. After that, volunteering to test the effectiveness of life jackets didn't seem so risky...

Pask's work remained secret until the war ended, when he wrote up his self-experiments and won the distinction of being the only man to have carried out all his research while asleep. But it was another 20 years before he owned up to one particular wartime investigation. He and his colleague John Gilson had been asked if there was any way that Winston Churchill could safely take oxygen at altitudes above 8,000 feet and still smoke a cigar. The pair took an ordinary cigar holder and added a side tube for the oxygen. Unfortunately, they couldn't find a valve that would ensure the oxygen always flowed in the right direction.

'The trouble was that the device worked unless you happened to put your tongue over the end of the cigar holder inside your mouth,' said Pask. 'This caused the oxygen to flow not into Churchill, but out through the lighted cigar. The wretched thing then burst into a brilliant white flame and about an inch of the best Havana disappeared before you could realise what had gone wrong.'

## ☀ A nose by any other name

*'The patient leaped upon a table and, laying himself on his back, with his head supported by a pillow, refused to be held; saying "I hope I shall behave like a man!"' The man on the table was one Captain Williamson, an officer in the British army. The man wielding the scalpel was London society surgeon Joseph Carpue. Both prayed the operation would be a great success. Williamson was desperate for a new nose; his own had been destroyed by disease. Carpue was equally eager; he had waited 20 years for a patient who was willing to submit to a new type of surgery. If all went well, the captain would be able to show his face in public once more and Carpue would win fame as a restorer of noses.*

Joseph Carpue had been pondering the possibilities of a new nose for years. He didn't want one himself, but there were plenty of people who through no fault of their own had lost their noses and needed new ones. The noseless were shunned, their disfigurement assumed to be a punishment for their sins. With a new nose they could take their place in polite society again. As one of the most brilliant surgeons in Georgian London, Carpue was the man for the job. He had the skills. He knew the theory. All he had to do was find someone willing to be his first patient.

Carpue's urge to create a new nose was prompted by the appearance of a letter in *The Gentleman's Magazine* on 9 October 1794. The letter, written by 'B. L.', a British army surgeon in India, described how the driver of a bullock cart, a man called Cowasjee, had lost his nose and acquired a new one. The replacement nose had been created from a flap of skin cut from his forehead and grafted onto the stump of the old one. It was a brilliantly simple piece of surgery. Carpue couldn't wait to try it.

People were fascinated by the bullock driver's story and soon his nose was the talk of the town. In 1792, Cowasjee had been carting supplies for the British army, then at war with the great Tippoo Sultan, ruler of Mysore. He was captured by Tippoo's men, who cut off one of his hands and his nose before sending him back to the British. 'For 12 months he remained without a nose, when a new one was put on by a man of the Brickmaker caste, near Poonah,' wrote B. L.

The brickmaker moulded a nose from wax, squashed it flat to create a template, and laid it on Cowasjee's forehead. Then he drew around the wax, cut along the line and lifted up the skin to create a thin flap, leaving a strip of skin, or pedicle, still firmly attached between the eyes. After paring away the scar tissue from the remains of the old nose, the brickmaker-surgeon twisted the flap over the stump, and inserted the edges into a series of neat incisions. The new nose was secured with slips of cloth soaked in a plant extract that dried to form a sort of cement. The pedicle kept blood flowing into the flap until the graft had taken, at which point it was snipped away. 'The artificial nose is secure and looks nearly as well as the natural one,' reported B. L.

Four years later, writer and traveller Thomas Pennant reported in *The View of Hindoostan* that Cowasjee's nose was bearing up well. It could 'sneeze smartly, distinguish good from bad smells, bear the most provoking lug, or being well blown without danger of falling into the handkerchief'.

While Cowasjee's story enthralled the British, in India it was not at all astonishing. As B. L. pointed out, the operation had been practised there 'from time immemorial'. No one should have been surprised that this early form of plastic surgery had been perfected in India. Noselessness was rife. Nasal amputation was a common punishment for thieves and adulterers, intended to humiliate and to deter others. The pressing need for new noses had led inevitably to the development of a highly effective technique for replacing them.

In late 18th-century Europe, nasal amputation was rare. Occasionally a man might have his nose sliced off in battle or during a duel, but disease was the more common destroyer of noses. Cancer, liver disease and syphilis all ate away at noses, producing a pool of potential patients in need of new ones. Inspired, Carpue set about learning more.

Building new noses this way had been common practice in India since around AD 1000, but there were even earlier techniques. Around 600 BC, an Indian surgeon called Sushruta wrote instructions for repairing noses using a flap of skin from the cheek. Skills developed in India spread west to ancient Greece and Rome, where many gladiators were grateful for them. The Byzantine emperor Justinian II had his nose amputated by his enemies during a coup in AD 695; he got himself a new nose and made a comeback nine years later. Replacement noses popped up again in 15th-century Sicily, where a family of barbers had a sideline in nasal surgery. And in the late 16th century another Italian, Gaspare Tagliacozzi, wrote a book outlining his version of the Italian nose job. He grafted skin from the inner arm, a procedure that required patients to sit for months with one arm strapped to their noses. This practice quickly died out: the church disapproved of it and there were malicious, but unfounded, rumours that a Tagliacozzi nose had a tendency to drop off. The ancient art of fixing noses vanished from Europe.

Carpue lived in a more enlightened age and resolved to revive the art of restoring noses. He, though, would do it the Indian way: it might be older but it had the advantage of simplicity and from what he had heard it was almost always successful.

If Carpue had expected a great rush for new noses, he was to be disappointed. He waited patiently, and in September 1814, after a wait of almost 20 years, Captain Williamson finally came along. He had heard that Carpue restored noses. He had suffered from liver disease and had been taking mercury for it for years, he said. The mercury had destroyed his nose. Could Carpue fix it?

That left Carpue in a quandary. He wasn't sure he wanted Williamson to be his first patient. After singing the praises of the Indian operation for so long, he had to get a good result. Williamson was not a well man. The mercury he had been prescribed had wrecked his constitution and might ruin his prospects of healing. And while he blamed his troubles on liver disease, the real culprit was syphilis. In India, amputation with a sharp blade left clean, neat stumps to work on. Syphilis was not so tidy: it would be risky to fix a new nose to such a diseased base. There was a good chance the graft wouldn't take.

On the pretext of preparing the captain's face for the operation, Carpue made incisions around the ruined nose. The wounds healed well. He decided to operate. On 23 October, after several test runs on corpses, he went to work on Williamson's nose. The captain was really not the best guinea pig. Apart from his syphilis, he had an unusually low forehead – too low to provide a nose-sized flap. Carpue would have to cut skin from the scalp too. What if the hair regrew along the base of the new nose? Would it push the nose off his face? He decided to risk it.

Following B. L.'s instructions, Carpue modelled a wax nose, flattened it, laid it on Williamson's head and drew around it with red paint. Then he began to cut, starting with the stump. First he made a cut beneath the nose to take the end of the flap and create a central septum. Next, he made cuts to the side of the nose to take the outer edges of the flap – what would become the wings of the nostrils. That done, he cut around the red line on the forehead, turned the flap around, fitted the edges into the cuts he had made, and stitched them in place. Finally, he stuffed pieces of lint into the nostrils to keep them open. The whole operation had taken 15 minutes.

Three days later, in the presence of one of Williamson's friends, Carpue removed the dressings. The friend, he wrote later, was amazed, shouting 'My God, there *is* a nose.' It was a nose, but not one Carpue liked the look of much. He was alarmed by its flatness. Then it swelled hideously. For weeks the nose alternated between balloon and pancake. But after four months, Carpue was satisfied: the nose had healed. There was one last thing to do. He snipped through the pedicle and tidied up the wound. The nose job had arrived.

# 2 Bad timing

Just how many times was the wheel reinvented? Undoubtedly lots – but most of the time no one noticed. Some inventors are simply too far ahead of their time; some miss the boat completely; while others get it just about right but can't stop tinkering with their baby…

## ☀ Dr Coley's famous fever

*The man's medical records were quite clear. His case was hopeless. In the space of three years, he had had five operations to remove a tumour from his neck. The last was a failure: it was impossible to remove the whole tumour. He would die soon. As if that wasn't bad enough, the poor man then suffered two attacks of erysipelas, a skin infection that produced a lurid red rash and high fever. But when the fever broke and the man recovered, his tumour had vanished. Seven years later, he was still alive and well. There could be only one explanation: whatever had caused the fever had also destroyed the cancer. William Coley, a young surgeon at the Memorial Sloan-Kettering Cancer Center in New York, reasoned that if a chance infection could make tumours vanish, then it should be possible to achieve the same effect by deliberately infecting patients.*

For centuries, the only real treatment for cancer was surgery. By the 1870s, when William Coley began his medical career in New York, patients at least had the benefit of antiseptics and effective, if rather dodgy, anaesthetics. Yet despite these improvements, surgeons were having less success than their predecessors a century earlier. Coley had become only too aware of this.

Coley's first cancer patient was a 19-year-old girl with a malignant tumour in her right arm. Despite early diagnosis and swift amputation of the arm, the cancer returned and spread quickly. She died soon after. Coley began delving into old medical books to find out more about the disease. In the past, he discovered, cancers that were cut out rarely returned. One surgeon working in the 1770s cured six out of every seven patients. Yet by the second half of the 19th century, surgery cured only one in four.

Coley discovered something else. For hundreds of years, doctors had reported cases of tumours that disappeared, apparently spontaneously. Searching his own hospital's records he came across a patient whose neck tumour had vanished seven years before. Coley tracked him down and found he was still alive. All these people had something in common apart from their miraculous recovery. They had all been struck down by an acute infectious disease. It might have been flu or measles, or something much worse – malaria, smallpox or syphilis, or like the man in New York, erysipelas. In most cases, when their fever subsided, their tumours had broken down and been absorbed or sloughed off.

Infection seemed to be the key to the 'miracle cures' and also to the success of early surgeons. Without antiseptics or antibiotics, patients who had their cancers carved out almost inevitably picked up infections from dirty instruments, dirty hands and unhygienic dressings. But by the 18th century, some surgeons knew enough about infection to try primitive forms of 'immunotherapy' on their cancer patients. Some slapped on a septic dressing from an infected patient. A few went so far as to inject material from patients with malaria or syphilis into tumours. Sometimes it worked. The infection seemed to reach those last vestiges of a tumour that the surgeon couldn't.

But by Coley's time, cleanliness and hygiene were the order of the day and surgeons had no truck with the idea of deliberately infecting patients. Occasionally, though, a patient caught an infection by accident. In the case of the man with the neck tumour, Coley had seen the result. 'There was no possibility of attributing the cure to any other cause than the erysipelas,' he told the New York Academy of Medicine in 1892. And if accidental infection with erysipelas could get rid of a tumour, he argued, 'it seemed fair to presume that the same benign action would be exerted in a similar case if erysipelas could be artificially produced'.

Erysipelas is caused by *Streptococcus pyogenes*, a bacterium that produces painful and unpleasant symptoms but which, unlike gangrene, syphilis or tuberculosis, was rarely dangerous. Coley decided to 'inoculate the first case of inoperable sarcoma that should present itself'. In May 1891, he found a willing volunteer. The man had tumours in both his neck and his tonsils, and despite recent surgery they had reappeared and were growing fast. Coley injected a streptococcal soup directly into the tumours, every day or two for the next two months. The tumours shrank. The man began to feel better. In August Coley stopped the injections and the tumours began to grow again.

Coley acquired a more potent culture of streptococci and tried again. This time the patient developed full-blown fever. 'The disease ran its course and I made little effort to check it,' reported Coley. 'At the end of two weeks, the tumour of the neck had disappeared.' Almost two years later, when Coley reported his results, the tumour in the neck had not returned. And although the second tumour hadn't shrunk, it hadn't grown either. 'Its malignant character must have been greatly modified,' reported Coley, 'as sarcoma of the tonsil is known to be rapidly fatal.'

Coley tried his treatment on more patients. His sixth case was memorable. The patient was a middle-aged cigar maker with a lumpy skin tumour on his back and a second tumour in his groin, this one the size of a goose egg. Surgery failed and both tumours soon grew back. Coley injected them with his streptococcal brew. They shrank but showed no sign of breaking down.

He tried again with a fresh culture acquired from the great German bacteriologist Robert Koch. Almost immediately, the cigar maker grew feverish. His temperature hit 40 °C. The lump on his back responded immediately. 'From the beginning of the attack the change that took place in the tumour was nothing short of marvellous,' wrote Coley. 'It lost its lustre and colour and had shrunk visibly in size within 24 hours.' A few days later, the second tumour began to break down too. 'Three weeks from the date of the attack both tumours had entirely disappeared.'

To start with, Coley believed he needed live bacteria. But even daily injections sometimes failed to produce fever, while in other patients the infection ran out of control. Coley decided that the key factors in his streptococcal soup were bacterial toxins – and that dead bacteria might work just as well. He finally settled on a mix of dead streptococci and another bacterium called *Serratia marcescens*, which became known as Coley's toxins. These had the advantage of triggering the symptoms of sickness – chills and fever – without actual infection. Coley insisted that it wasn't so much the identity of the bacteria that was important but the technique he used to treat patients. It was essential to inject the toxins deep into the tumour, as often as necessary to cause fever, and keep this up for weeks or even months.

Coley's toxins produced good results from the start. Patients thought to be beyond all help saw their cancer disappear. Many of those who weren't cured lived much longer than they might otherwise. 'He had successes you simply couldn't hope for today, curing even extensive metastatic disease,' says Stephen Hoption Cann, an epidemiologist at the University of British Columbia.

Although patients flocked to the hospital for treatment,

Coley found it increasingly difficult to treat them. His boss thought another new treatment, radiotherapy, more promising. 'The response to radiotherapy was highly predictable,' says Hoption Cann. 'Irradiate the tumour and it shrinks. Unfortunately, it always came back.' Its supporters believed that with a few modifications, radiotherapy would eventually be able to cure cancer. 'Yet, here we are a hundred years later and if the cancer has spread, radiotherapy is not curative.'

Coley successfully treated hundreds of patients but his treatment was trickier and took longer. The dose had to be tailored to each patient and gradually increased to keep the immune response going. After his death in 1936, interest in Coley's toxins waned. Radiotherapy and, later, chemotherapy became standard treatments. Both knock out the immune system and so infection became something to be avoided at all cost.

The search for ways to stimulate the immune system to fight cancer continues. Researchers have narrowed their search, focusing on ways to trigger production of specific types of anti-cancer cells or particular tumour-suppressing molecules.

But as Hoption Cann points out, Coley's vaccine worked precisely because it was so crude and stimulated a general immune response. More important, he says, the immune system works at its best during a fever. 'The body produces more immune cells. They are more mobile and more destructive.'

What's more, there is growing evidence that the more infections people have – especially if they develop fever – the less likely they are to suffer certain cancers. Or, as the great 17th-century physician Thomas Sydenham once said: 'Fever is a mighty engine which Nature brings into the world for the conquest of her enemies.'

# ☀Sunshine for sale

*'It has been a favorite pastime for the dreary gentlemen who juggle with statistics, solemnly to calculate the date on which we shall all freeze to death from exhaustion of the coal supply,'* Harper's Weekly *noted dryly in 1903. It was not a new worry even then, for Britain's geologists had been summoned to Parliament as early as 1829 to estimate when the country's coalfields would be mined out. Victorian engineers found solar power an irresistible solution, and one of the most eminent, John Ericsson – inventor of the screw propeller and the ironclad warship USS* Monitor *– devoted the rest of his life to it. 'A couple of thousand years dropped in the ocean of time will completely exhaust the coalfields of Europe,' he remarked. 'The application of the solar engine is almost beyond computation, while the source of its power is boundless.' The result? In 1914, his ideas inspired a scheme to run the whole of Europe on solar power, and it might have even worked…*

Though fly-by-night solar firms flourished in America in the late 19th century, their products were not entirely fanciful: some irrigation systems and hot water tanks ran on solar motors. But most solar engineers believed the technology's future lay in Africa, where French inventor Augustin Mouchot had built desalination plants in Algeria in 1877. His rival, Swedish-American John Ericsson, foresaw a day when energy politics would shift the balance of power towards the deserts of the Middle East: 'The rapid exhaustion of the European coal-fields will soon cause great changes in reference to international relations, in favor of those countries which are in possession of continuous sun-power. Upper Egypt, for instance…' It was a prediction that one inventor took quite literally.

Frank Shuman started small. In 1906 he built a 'hot box', a blackened box under heat-trapping glass. Attached to it, reported *Engineering News*, was 'a tiny toy engine such as sold for a dollar'. He quickly moved up to a larger engine and bigger hot box. His third effort, in 1907, took up the whole of his backyard in Tacony, a suburb of Philadelphia. This time, the solar collector was a network of blackened pipes covering an area of 96 square metres and filled with ether, which had a conveniently low boiling point. The heated vapour drove a water pump.

During the summer, neighbours could watch the 2.2-kilowatt contraption pump thousands of barrels of water a day. It even ran during the winter, albeit more slowly. It was certainly safer for neighbours to gawk at than the sun-motors designed by Mouchot and Ericsson. These used costly parabolic mirrors to focus sunlight on a boiler assembly, sometimes producing a temperature of more than 1,000 °C. But Shuman's engine was not perfect either. It cooled off so much overnight, and took so long to get going each day, that a reporter wondered aloud 'if it belonged to the Sun union'.

Shuman eventually worked out how to prevent much of the heat loss by floating a thin layer of paraffin on water in solar collectors to absorb and trap heat. He then lined the bottoms of the collectors with macadam as a watertight insulation that was, he pointed out, 'black all through, and therefore will never require painting'. Shuman was particularly concerned with durability and simplicity and felt asphalt would also be 'practically everlasting'. Furthermore, he addressed one of the inherent problems of solar energy: how to keep it flowing when the sun wasn't out. Like most other solar engines, his pumped hot water up to an insulated tank, from where it would drive a turbine later.

Shuman's designs were only a marginal improvement over those of his rivals, but that didn't matter. What Shuman brought to the field was a forceful personality and brilliant sales patter. He made solar power seem inevitable and persuaded investors they would lose out if they weren't in at the start of something this big. Soon he was issuing stock for The Sun Power Company and, after an encouraging write-up in *Engineering News,* he won backing from a group of British investors. Along with this financing came two engineering partners, Alfred Ackerman and Charles Boys.

At first they drew up plans to build plants in Florida and Arizona. But Shuman knew there were bigger prizes. In 1912, he and his crew packed their bags and set off for the ultimate destination of any solar engineer: North Africa. Shuman's vision took him to Meadi, a farming town about 25 kilometres from Cairo, where the Sun Power team built the largest solar power plant the world had ever seen.

When it was finished in July 1913, the plant had more than 1,200 square metres of mirrored V-shaped collectors focused on a blackened pipe. This boiled water to power a 40-kilowatt irrigation system. The Meadi plant ran 24 hours a day, and it was built of simple but strong materials. 'Any common engineer can run it,' noted *Scientific American.* 'Owing to the fact that ordinary materials are used in its construction, repairs can be made easily and everything is above ground and readily accessible.'

One visitor to Meadi was Lord Kitchener, the Consul General of Egypt. Impressed with the pumps operating by the ghostly power of the sun, he offered Shuman a 120-square-kilometre plantation in the Sudan to build his next system. Not to be outdone, the Germans called a special session of the Reichstag to host Shuman. They too wanted solar plants in their African territories, and dangled $200,000 before Shuman to build one.

Shuman was jubilant. Solar power, he boasted, was now inevitable: 'There is not a single "should give" or "guess" about it. Sun power is now a fact, and is no longer in the "beautiful possibility" stage. It can compete profitably with coal in the true tropics now.' He had more than irrigation in mind for his future: why not spread the plants across Saharan Africa to generate electricity for all of Europe?

Shuman's vision was laid before readers of *Scientific American* in February 1914: 'I have taken as a basis the figure of two hundred seventy million horse-power [200 gigawatts] continuously throughout the year being equal to all the coal and oil mined during the year 1909 throughout the world... Taking the actual work of our plant as a basis, it would only be necessary to cover 20,250 square miles [52,500 square kilometres] of ground in the Sahara Desert with our sun heat absorber unit... Surely from this showing, the human race can see that solar power can take care of them for all time to come.'

Endless power, clean and guaranteed, and financed, Shuman suggested, by spreading the cost with bond issues. That cost would be staggering – 'ninety-eight odd billion dollars' by Shuman's estimate. But, he said, 'this vast investment would not be made for or by the individual, but for and by the entire human race'. The scale of Shuman's vision was awesome but, as he pointed out, the technology was there. All it needed was the will to build it. And Shuman was alert to the political and environmental havoc that petrochemicals might visit upon the world. 'One thing I feel sure of,' Shuman insisted, 'and that is that the human race must utilise direct Sun power or revert to barbarism.'

Humanity reverted to barbarism sooner than Shuman imagined, and he did not live to see the end of the First World War. Kitchener went down on the ill-fated HMS *Hampshire*, the Meadi team was assigned to war duties, and all British and German plans for solar power were swept aside. With their post-war economies gutted, and oil in the ascendancy, the notion of spending billions on sun power was unthinkable. The vision of a solar-powered Europe sank back into darkness.

# ☀The chunkiest chip

*Today they are everywhere. Production lines controlled by comput-*
*ers and operated by robots. There's no chatter of assembly workers,*
*just the whirr and click of machines. In the mid-1940s, the worker-*
*less factory was still the stuff of science fiction. There were no com-*
*puters to speak of and electronics was primitive. Yet hidden away in*
*the English countryside was a highly automated production line*
*called ECME, which could turn out 1,500 radio receivers a day with*
*almost no help from human hands.*

The Indian government ordered 20,000 of them. China's president, Chiang Kai-shek, bought 25,000 and might have
ordered more if the People's Revolution hadn't disrupted his
plans. By 1948, John Sargrove's radios were selling like hot
cakes in Asia and the Far East. Which was exactly what he
had intended when he designed the world's first automatic
assembly line. The line turned out radios so cheaply that
people in some of the world's poorest nations could afford to
buy them.

For more than a decade, Sargrove had been trying to
figure out how to make cheaper radios. Automating the
manufacturing process would help. But radios didn't lend
themselves to such methods: there were too many parts to fit
together and too many wires to solder. Even a simple receiver
might have 30 separate components and 80 hand-soldered
connections. At every stage, things had to be tested and
inspected. Making radios required highly skilled labour –
and lots of it.

In 1944, Sargrove came up with the answer. His solution
was to dispense with most of the fiddly bits by inventing a
primitive chip – a slab of Bakelite with all the receiver's
electrical components and connections embedded in it. This
was something that could be made by machines, and he
designed those too. At the end of the war, Sargrove built an

automatic production line, which he called ECME (electronic circuit-making equipment), in a small factory in Surrey.

When Sargrove unveiled his invention at a meeting of the British Institution of Radio Engineers in February 1947, the assembled engineers were impressed. So was the man from *The Times*. ECME, he reported the following day, 'produces almost without human labour, a complete radio receiving set. This new method of production can be equally well applied to television and other forms of electronic apparatus.'

From the outside, ECME looked like a very ordinary set of metal cabinets, 20 metres long and with doors at regular intervals. The cabinets housed a series of interconnected machines that built up the circuits.

The starting point was the piece of Bakelite, moulded with a pattern of grooves and depressions on each side. When these were filled with metal, they formed all the conductors, inductors, capacitors, resistors and so on that the receiver needed, all connected in exactly the right way. An operator sat at one end of each ECME line, feeding in the plates. She didn't need much skill, only quick hands. From now on, everything was controlled by electronic switches and relays.

First stop was the sandblaster, which roughened the surface of the plastic so that molten metal would stick to it. The plates were then cleaned to remove any traces of grit. The machine automatically checked that the surface was rough enough before sending the plate to the spraying section. There, eight nozzles rotated into position and sprayed molten zinc over both sides of the plate. Again, the nozzles only began to spray when a plate was in place.

The plate whizzed on. The next stop was the milling machine, which ground away the surface layer of metal to leave the circuit and other components in the grooves and recesses. Now the plate was a composite of metal and plastic. It sped on to be lacquered and have its circuits tested. By the time it emerged from the end of the line, robot hands had fitted it with sockets to attach components such as valves and loudspeakers. When ECME was working flat out, the whole process took 20 seconds.

ECME was astonishingly advanced. Electronic eyes, photocells that generated a small current when a panel arrived, triggered each step in the operation, so avoiding excessive wear and tear on the machinery. The plates were automatically tested at each stage as they moved along the conveyor. And if more than two plates in succession were duds, the machines were automatically adjusted – or if necessary halted. In a conventional factory, workers would test faulty circuits and repair them. But Sargrove's assembly line produced circuits so cheaply they just threw away the faulty ones.

Sargrove's circuit board was even more astonishing for the time. It pre-dated the more familiar printed circuit, with wiring printed on a board, yet was more sophisticated. Its built-in components made it more like a modern chip.

The receivers had many advantages over their predecessors. With fewer components they were more robust. Robots didn't make the sorts of mistakes human assembly workers sometimes did. 'Wiring mistakes just cannot happen,' wrote Sargrove. No wires also meant the radios were lighter and cheaper to ship abroad. And with no soldered wires to come unstuck, the radios were more reliable. Sargrove pointed out that the circuit boards didn't have to be flat. They could be curved, opening up the prospect of building the electronics into the cabinet of Bakelite radios.

Sargrove was all for introducing this type of automation to other products. It could be used to make more complex electronic equipment than radios, he argued. And even if only part of a manufacturing process were automated, the savings would be substantial.

But while his invention was brilliant, his timing was bad. ECME was too advanced for its own good. It was only competitive on huge production runs because each new job meant retooling the machines. But disruption was frequent.

Sophisticated as it was, ECME still depended on old-fashioned electromechanical relays and valves – which failed with monotonous regularity. The state of Britain's economy added to Sargrove's troubles. Production was dogged by power cuts and post-war shortages of materials. Sargrove's financial backers began to get cold feet.

There was another problem that Sargrove hadn't foreseen. One of ECME's biggest advantages – the savings on the cost of labour – also accelerated its downfall. Sargrove saw automation as the way to solve post-war labour shortages. With somewhat Utopian idealism, he imagined his new technology would free people from boring, repetitive jobs on the production line and allow them to do more interesting work. 'Don't get the idea that we are out to rob people of their jobs,' he told the *Daily Mirror*. 'Our task is to liberate men and women from being slaves of machines.' He travelled the country giving talks on the advantages of automation. It wouldn't cause unemployment, he argued. 'It means redeployment – to relieve people of monotonous jobs so that they can do more interesting work beyond the capacity of a machine.'

The workers saw things differently. They viewed automation in the same light as the everlasting light bulb or the suit that never wears out – as a threat to people's livelihoods. If automation spread, they wouldn't be released to do more exciting jobs, they'd be released to join the dole queue. Financial backing for ECME fizzled out. The money dried up. And Britain lost its lead in a technology that would transform industry just a few years later.

# ☀️Mark Twain's big mistake

*On 5 January 1889, Mark Twain watched with glee as an extra-ordinary invention clattered into life. 'At 12.20 this afternoon,' wrote America's most famous novelist, 'a line of moveable type was spaced and justified by machinery, for the first time in the history of the world! And I was there to see it. It was done automatically, instantly, perfectly.' Surely this new printing press, into which he had poured his royalties from* The Adventures of Huckleberry Finn, *would launch a revolutionary new information age. 'Tele-phones, telegraphs, locomotives, cotton gins, sewing machines, Babbage machines, Jacquard looms, perfecting presses, Arkwright's frames – all mere toys, simplicities!' The Paige Compositor should have made Twain's fortune. Instead, it nearly destroyed him.*

In Mark Twain's time, publishers were stuck in a technologi-cal bottleneck. The invention in 1843 of both mass-produced wood-pulp paper and the steam-powered rotary press allowed printing at blinding speeds and in huge quantities, yet typesetting remained a stubbornly slow process that had barely changed in centuries. Type was still set by hand, with a skilled printer preparing approximately 800 'ems' (about 100 words) an hour.

But by the early 1870s James Paige, a gifted mechanic in Rochester, New York, began to design the Paige Compositor, a clever mechanical rendering of the movements of a human typesetter. When Twain was introduced to Paige in 1880, he was immediately dazzled: 'He is the Shakespeare of mechanical invention,' pronounced the novelist. As a publisher and former printing apprentice himself, Twain was convinced the machine would revolutionise the industry. By his own calculations, the Paige Compositor might be worth $150 million over the lifetime of its patents. He immediately began to pour money into Paige's project.

In December 1881, Paige demonstrated a prototype to a

gathering of newspapermen and publishers. It was big, standing almost 2 metres high, and it was fast, enabling a single typesetter to set words from a keyboard at speeds of up to 12,000 ems an hour. Unfortunately, Paige kept pausing to tinker with his baby, leaving the audience staring at motionless machinery for most of the time. Even so, W. D. Howells, editor of *The Atlantic Monthly*, was impressed. 'It did everything but walk and talk,' he declared.

Indeed, the Paige Compositor was a wondrous piece of work... when it worked. The machine was temperamental: a bit of broken or transposed type quickly brought it to a halt. Paige kept thinking up further improvements, and the compositor grew dizzyingly complex. Eventually it had 18,000 parts and 800 shaft bearings. Undeterred, and even though Paige ignored his pleas to show investors a working model, Twain continued to hand over large sums of money.

What captivated Twain? The prospect of riches, certainly. But Paige also envisioned something that writers and booksellers still hanker after: print on demand. An article in the *Chicago Tribune* in 1892 retains a curiously familiar ring. 'The poor but ambitious author – Mr. Paige says he is everywhere – has written a work which he feels within his burning soul is sure to be the long-looked for American novel. It is brief but it bristles. His pockets are empty, of course, and his frequent trips to heartless publishers have left him nothing but worn shoes and his manuscripts. He hears of this machine; he goes to it; his book is to be of 200 pages; he borrows $5 from his landlady and presto, in twenty minutes he has the prized novel, the child of his brain, in cold but clear type in his inside pocket.'

It remains an alluring vision. But then Twain always had a remarkable capacity for wonder, delight and sometimes sheer gullibility. In his youth he had briefly prospected for gold in Nevada, and his fascination with get-rich-quick schemes never left him. He sank thousands of dollars into the Fredonia Watch Company, which produced a 'Mark Twain watch', but proved curiously incapable of producing a Mark Twain royalty cheque.

Ever the optimist, he ploughed money into the Kaolotype, 'a chalk engraving process'. This too ate money with no apparent result, whereupon a company official assured Twain that 'there is one thing however you can depend upon. You are not being cheated and stolen from' – which, of course, he was. He then invested in a 'bed clamp' to prevent babies kicking off their blankets: it didn't work. In the midst of writing *Huckleberry Finn*, Twain even dabbled in a 'hand grenade' fire extinguisher, a glass bottle filled with flame retardant to be hurled into a fire. It didn't put out many fires, but it did burn a large wad of his cash.

Above all else, Twain invested in the Paige Compositor. One year after another passed without a completed machine, yet by 1886 Twain was handing over a ruinously generous $7,000 a month. Paige had troubles of his own. There were wild rumours that he had become a millionaire, attracting the inevitable spongers and con artists.

An actress in Chicago sued Paige for breach of promise – and $800,000 – claiming that the bewildered inventor had promised to marry her. Fortunately for Paige, the case fell apart when her husband showed up and contradicted her.

Paige laboured on, insisting on ever more and complicated improvements until even Twain grew alarmed. 'Business sanity would have said put it on the market as it was, secure the field, and add in improvements later,' wrote Twain. Instead, Paige kept dithering, until finally in 1886 disaster struck. Rival inventor Ottmar Mergenthaler unveiled his Linotype, a less ambitious but vastly more practical and reliable typesetting workhorse. Even now Paige continued to perfect his grand invention; years of maddening delay were relieved only by fleetingly successful demonstrations, such as the one in 1889 that had Twain waxing so lyrical. Paige's endless stalling exasperated his backer. 'If he were drowning,' said Twain, 'I would throw him an anvil.'

In a desperate attempt to recoup something from his investment before the Linotype wrecked any chance of it, Twain's life became an endless cycle of more bills from Paige, more missed deadlines and more humiliating attempts to summon up some new investors. A few friends and relatives pitched in, including a then-unknown theatrical manager by the name of Bram Stoker. Hard-nosed venture capitalists were less willing to take the risk.

The years rolled on and still no working model emerged from Paige's workshop. With each new deadline, Twain pinned his hopes of financial salvation on Paige's rapidly receding dream. 'I have never been so desperate in my life,' Twain privately admitted in 1893, 'and for good reason, for I haven't got a penny to my name.'

A single working model was finally produced for a trial run at the *Chicago Tribune* in 1894, but by then it was far too late for Paige's temperamental machine. At least $150,000 of Twain's money was gone. In the end, Paige had to shut up shop, and America's most popular author sold his Hartford mansion and set off on an exhausting round-the-world lecture tour to pay off his creditors.

Paige had a harder time. After disappearing from public view, the inventor surfaced penniless at Chicago's Cook County Almshouse in 1917. He died a few months later and was buried in a pauper's grave. The machine on which he squandered endless time and money was bought as a prized relic by the Mergenthaler Linotype Company, the very competitor that had sealed its fate. Eventually it was donated to the Mark Twain House in Hartford, where it has never been taken apart, out of the eminently sensible concern that the only man who could put it back together died nearly a century ago. Mark Twain's glorious and doomed machine stands mute and unusable in the very house whose proud owner it drove out.

# ☀ Live from the Paris Opera

*French novelist Marcel Proust was sickly and often bedridden. By February 1911, he had one constant companion in the solitary confines of his cork-lined Paris sanctuary: a mysterious contraption that he kept by his bed. When he wasn't labouring over his magnum opus,* A la Recherche du Temps Perdu, *Proust would collapse into bed, grasp a pair of wires trailing into a primitive headset, and lose himself in Claude Debussy's opera* Pelléas et Mélisande. *What he heard was no scratchy gramophone recording, but a live broadcast – and in stereophonic sound.*

Parisians who stepped off the custom-built electric tram at the International Exposition of Electricity in 1881 could have been forgiven for believing the future had just arrived. The exhibition boasted eight apartments equipped with everything from electrically operated kitchens to a billiards room with an electric scoring system. But the star attraction was altogether more exotic: crowds queued behind 20 telephone receivers to hear a live performance from the Paris Opera. Inventor Clement Ader had placed dozens of microphones among the Opera's footlights and run telephone cable through the Paris sewers to the exhibition hall, delivering music into the ears of transfixed listeners.

The idea of broadcasting music is as old as telephony itself. On 22 March 1876, just 12 days after Alexander Graham Bell's momentous 'Watson, come here, I want you' ushered in the telephone era, *The New York Times* was predicting a wired future where multichannel entertainment was a household utility like gas and water. 'By means of this remarkable instrument, a man can have the Italian opera, the Federal Congress, and his favorite preacher laid on in his own house,' the paper proclaimed. 'Fifty eminent preachers, of different denomination, can be kept constantly on draught.' Within a year, pianist Frederick Boscovitz had phoned his rendition of

*Yankee Doodle* from Philadelphia to a rapt audience in Washington DC, and telephone enthusiasts elsewhere quickly followed suit.

But Ader's exhibit was different: Parisians were asked to place a pair of receivers over both ears for what was dubbed 'binauricular audition'. He had stumbled on the concept after hearing two microphones together on a pair of receivers. The result, reported *Scientific American*, was that 'singers place themselves, in the mind of the listener, at a fixed distance, some to the right and others to the left. It is easy to follow their movements, and to indicate exactly, each time they change their position, the imaginary distance at which they appear to be.' Exposition-goers were hearing stereo sound broadcast for the first time.

Ader and other would-be broadcasters were dogged by technical difficulties: the music came through faintly, and amplification technology was so rudimentary that the quality of the signal quickly degraded, even with only a few listeners on the line. 'Prima donna by telephone is not, as a rule, so satisfactory,' grumbled one writer for London's *Daily News*. There were problems placing the microphones in theatres too: whenever the music swelled, the primitive transmitters shook in their housings, and they had to be hung from rubber bands to isolate them from vibrations. The instruments of the orchestra presented their own challenge, and one early engineering note directed that the microphones should 'not be fitted in close proximity to the bass drum or the trombone'.

But by 1890, the Theatrophone home stereo system was ready to make its debut in Paris. For a hefty annual fee of 180 francs – equivalent to about three months' rent for a comfortable Paris apartment – plus 15 francs per performance, subscribers received a phone box with a headset and a transmitter so they could tell a Theatrophone operator which venue they wished to listen in to. For most Parisians, though, the Theatrophone was far too expensive to install at home. For their entertainment, coin-operated Theatrophone listening stations were installed in hotel lobbies and cafes across Paris; 50 centimes bought two and a half minutes of listening time. Curiously, the company found that one of its most popular offerings wasn't even the shows; it was the player-piano music that ran between performances. 'The artistry was not important,' recalled one customer. 'It was new, and that was enough.'

Interest quickly spread across the Channel to England. In 1892, crowds at the Crystal Palace pleasure grounds were treated to a demonstration line from the Lyric Theatre, and in 1894 Electrophone Ltd launched a London service. For £10 a year you could purchase four receivers and a multiheadset 'electrophone table' for family listening – and service direct from 18 London theatres. London churches were also wired for sound with microphones hidden inside a dummy Bible on the pulpit. By the turn of the century, the service had 600 well-heeled subscribers, including Queen Victoria.

Despite the US's pioneering role in musical telephones and a short-lived service based in New Jersey, numerous broadcasting schemes failed to get off the ground. Yet they found success in other countries, most notably Italy and Hungary. Driven by the prolific Hungarian inventor Tivadar Puskas, the Telefon Hirmondo service became Budapest's byword for news, music and entertainment. Beginning in 1893 and eventually building up to 15,000 subscribers, its receivers were a familiar presence at the city's public establishments. Hirmondo's broadcast schedule, with time checks, news and sporting updates every 15 or 30 minutes, and long musical performances in the evening, proved remarkably prescient of later radio programming.

Still, some artists were worried that listeners at home might empty the theatres. One of the final acts of Giuseppe Verdi's career was to sue a Belgian telephone music service in 1899 for broadcasting his opera *Rigoletto*. And private Theatrophone services were regarded by many as a plaything for the rich and idle. *Harper's Weekly* joked that Telefon

Hirmondo had made the city of Budapest 'the finest for illiterate, blind, bedridden and incurably lazy people in the world'.

French novelist Marcel Proust could certainly vouch for the bedridden, as he constantly requested opera programmes from his bedside Theatrophone. These broadcasts were one of his great solaces as he poured his strength into *A la Recherche du Temps Perdu*, even if the sound quality was sometimes so poor that, as Proust admitted in 1911, he had once mistaken a noisy crowd for a song: 'I thought the rumblings I heard agreeable if a trifle amorphous until I suddenly realised it was the interval!'

Even as it improved its sound and gained subscribers, the health of telephone entertainment was as precarious as Proust's. True, the invention of the vacuum tube dramatically increased the strength of the signal, allowing large numbers of people to share the same system. By 1914, the Hotel Wagram in Paris boasted a Theatrophone service in all 200 of its rooms. As the number of subscribers grew, prices fell and London's Electrophone service eventually had more than 2,000 customers.

But with the advent of cheap and portable wireless broadcasting, as early as 1923 *The Times* in London noted that radio broadcasting 'seems likely to seriously invade the province of the Electrophone'. Promoters dismissed such concerns, arguing that 'it would be a long time before broadcasting by wireless of entertainments and church services attained the degree of perfection now achieved by the Electrophone'. In fact, it took less than a decade. Telefon Hirmondo disappeared amid mergers, and by 1932 both the Electrophone and Theatrophone lines had gone dead.

For the next four decades consumers could reflect ruefully that their grandparents had stereo sound and they didn't. Listeners would not catch up again until the roll-out of FM stereo broadcasting in 1961.

# 3 Who dares wins

It's one thing to have a bright idea, it's quite another to profit from it. First, there has to be the opportunity – a gap in the market, a queue of potential clients. Then you have to grab it and hang onto it by fair means or foul.

## ☀ Henry's little pot of gold

*It all began with some chrysanthemums. In the late 1830s, Henry Bessemer was making his start in London by creating innovative metal dies and embossing devices. But on a visit to his sister, he found his talent for calligraphy called upon instead. She wanted him to write 'Studies of Flowers from Nature, by Miss Bessemer' on her portfolio of watercolour paintings of chrysanthemums and other flowers from the garden. And so Henry dutifully set about making gold ink to complete the lettering. To do this, he would have to obtain 'gold powder' from Mr Clark, the local art supplies merchant. Bessemer soon discovered that he was being more generous to his sister than he had anticipated: the gold powder cost an extortionate seven shillings an ounce. He doubted the powder contained actual gold, and yet it was far more expensive than raw brass. He was determined to get to the bottom of it.*

The more Henry Bessemer thought about it, the more perplexed he became. Just how could a tiny pot of faux gold cost so much? To determine what exactly was in the powder,

Bessemer created a solution of the art shop's powder and added dilute sulphuric acid to precipitate any actual gold. He watched impatiently: nothing happened. 'It is, probably, only a better sort of brass,' he reasoned. But on inquiring with art suppliers, none could explain the powder's origin. They only knew that it was made in Nuremberg, a centre of the brass trade, and that the manufacturing process was kept secret.

Bessemer quickly sensed a fortune to be made: 'Here was powdered brass selling retail at £5 12s. per pound,' he recalled in his autobiography, 'while the raw material from which it was made cost probably no more than sixpence.' He deduced that labour might account for the inflated price, and a visit to the reading room at the British Museum proved him right. Dipping into *De Diversis Artibus*, a 12th-century encyclopedia of craftsmanship written by the German monk Theophilus, he found a description of the method employed in Nuremberg: brass was pounded to leaf, then ground with a pestle and mortar, all the while being mixed with honey to prevent the particles from clumping. This glop needed repeated washings in hot water to remove the honey – a slow and laborious process, and an anomaly in the industrial age. What if, Bessemer wondered, he could manufacture it with steam power?

Bessemer tried using brass discs, ridged along their edges like a coin, and turning them against a lathe to throw off minute particles. His first tests were not promising: the powder had none of the lustre of the stuff he had bought in the art shop. Discouraged, he gave the project little more thought until a year later, when an acquaintance recounted a powder problem of a different sort. The acquaintance suspected a local merchant had sold him arrowroot adulterated with starch, and exposed the cheat by examining the powder under a microscope, discovering two distinct varieties of particles. Intrigued by this example, Bessemer compared the Nuremberg powder and his own under a microscope and 'saw in a moment the cause of my failure'.

Powder made from brass leaf presented flat, paper-like particles that created a unified surface, while Bessemer's particles, though similar to the naked eye, were rough and curled-up, scattering light haphazardly and with little lustre.

Sensing that he was now on the right path, Bessemer spent months secretly toiling in his workshop. His new solution was to flatten the brass grindings by passing them through steel rollers. He then ran them through a tumbler assembly, where the friction of particles cascading over each other would polish them to a pleasing shine. Viewing a sample of the resulting powder, one importer offered Bessemer £500 a year for use of whatever machinery had produced it – whereupon Bessemer knew his invention had to be worth a great deal more.

To make his fortune required the utmost secrecy, even from the Patent Office. He knew it would give the game away 'if all the details of my system were shown and described in a patent blue-book, which anyone could buy for a six pence'. Piracy was rife. Thomas Edison later complained that 'you cannot do anything in court for five or six years, and the infringer knows this'. But, he added: 'A trade secret is of value in the chemical line, for there it can be guarded.' Machines could be reverse-engineered or copied from patent papers; unpatented powder could not. To create a lucrative monopoly, Bessemer had to keep his secret... but how?

His solution was ingenious. 'There were powerful machines of many tons in weight to be made; some of them were necessarily very complicated, and somebody must know for whom they were... [So] when I had thus devised and settled every machine as a whole, I undertook to dissect it and make separate drawings of each part, accurately figured for dimensions, and to take these separate parts of the several machines and get them made: some in Manchester, some in Glasgow, some in Liverpool, and some in London, so that no engineer could ever guess what these parts of machines were intended to be used for.'

Working with his three trusted brothers-in-law, by 1841 Bessemer had converted a building in north London into a factory. With block and tackle, the four men were able to move massive machinery into place in secret and without any outside assistance. The completed factory was a wonder of paranoiac design: it had only one entrance and no windows, and the machinery was divided between three compartments, with drive shafts passing blindly through holes in the wall. It ran automatically with a minimum of oversight, and no one could see more than a fraction of the interior. If they had, they would have found remarkable contrasts. In one room, grinding machinery ran with 'the screech of a hundred discordant fiddles', while in the next, powder was blown onto a cloth-covered table, where particles separated into grades of fineness by the distance they travelled. 'It is difficult to imagine the beauty of this golden snowdrift of 40 ft in length,' Bessemer recalled.

Once restricted to higher-quality furniture and picture frames, mass-produced faux gold now came within everyone's reach: gold picture frames, bronzed plaster statuary, and gold inks proliferated. By experimenting with different alloys and copper suppliers – he was particularly fond of melting down barrels of Russian kopeks – Bessemer also created an array of coloured powders. Yet the product was not perfect. As they aged, oxidation took its toll, rendering them increasingly dark and unattractive. Aware of this, Bessemer concocted a gold paint containing calcium succinate to slow oxidation, sold cheaply in paired ready-to-mix bottles of powder and medium.

Modern gold-coloured paints are made from mica coated with titanium oxide, which avoids the oxidation that bedevilled brass powder. But in the Gilded Age, it was Bessemer's brass – or the more elegant sounding 'bronze' it soon became – that provided the gilding. And the only way to get it was from his mysterious, windowless building.

Despite the powder's popularity, Bessemer managed to

keep his manufacturing process secret for more than 40 years. The steady income from his business freed him to pursue other inventions, and he went on to win 114 remarkably innovative and profitable patents – including the one he is most remembered for, the process that revolutionised steel-making. Not bad for a venture that started with a few watercolour paintings of chrysanthemums.

## ☀ Hens' eggs and snail shells

*The Hippocratic oath includes this curious promise: 'I will not cut for stone… I will leave this operation to be performed by practitioners.' Bladder stones are among humanity's oldest known ailments, and its surgery an ancient procedure – but, as Hippocrates warned, one so dangerous that it should not be trusted to doctors. Only specialists could operate to relieve this agonising affliction. In an age before anaesthesia and antiseptics, lithotomies were often fatal. Samuel Pepys was so glad to survive his that he placed his stone in a reliquary and threw a party for it every year. Others set their stones into jewellery and printed cards commemorating their surgery. Those less determined to defy death resorted to quack medicine or even scraping their insides with wire. Benjamin Franklin, as usual, outdid everyone. When stones blocked his urethra, he dislodged them by standing on his head and urinating upside down.*

Benjamin Franklin had plenty of company in his medical miseries: Isaac Newton also suffered from bladder stones, as did Peter the Great of Russia and Napoleon Bonaparte. Bladder and kidney stones were once a far more common affliction than they are today. And while some sufferers passed their stones relatively painlessly, others were less fortunate, suffering excruciating pain, vomiting, fever, bloody urine and permanent kidney damage.

So readers sat up and took notice when on 27 April 1738

*The Gentleman's Magazine* published this brazen offer by one Joanna Stephens: 'Mrs Stephens has proposed to make her Medicine for the Stone publick, on Consideration of the Sum of £5,000 to be lodged with Mr Drummond, Banker.' It was an immense sum for an obscure individual with no medical degree – and a woman to boot. Were sufferers really that desperate? They were indeed: more than £1,000 poured into the bank.

It helped that the prominent doctor and philosopher David Hartley vouched for the mysterious Mrs Stephens. He spoke from experience: her concoctions had cured him of his own stones, he claimed. Exhorting the public to accede to her demand for £5,000, pointing out that the money could be held by a third party until trials had shown her cure worked, he wrote: 'I therefore perswade myself, that Mrs Stephens will appear to you in a different light from common Pretenders to NOSTRUMS…'

Hartley campaigned relentlessly on Stephens's behalf, collecting 155 accounts of patients cured by her secret medication. But in 1739, with the collection stuck at £1,352 and 3 shillings, she made her boldest move: she petitioned parliament for the full £5,000. Her novel proposal was that she would publish the secret ingredients, allowing everyone to try it for themselves. If the cure worked, a grateful nation would make her rich. Parliament approved her extraordinary request, and Stephens turned over her recipe to the Archbishop of Canterbury.

On 16 June 1739, a full-page notice appeared in *The London Gazette*. Doctors and stone sufferers rushed into their gardens for ingredients. 'Take Hens Eggs well drained from the Whites, dry and clean,' Stephens instructed. 'Crush them small with the Hands, and fill a Crucible… ' They then added crushed snail shells, slaking and baking the mixture into a fine powder; to this was added soap and honey, as well as chamomile, fennel, parsley, burdock and burnt swine's-cress.

Stephens's day of reckoning came soon enough. On 5 March 1740, four selected stone patients were summoned to the House of Lords to appear before a panel that included the Archbishop of Canterbury, the speaker of the House of Commons, the president of the Royal College of Physicians and the Prince of Wales's personal surgeon. The first to testify, a Mr Gardiner, said that after eight months of treatment his symptoms had vanished, and he had passed his stones. The next three men all gave similar accounts: after taking Stephens's medicine their urine became smelly, turned cloudy, and then – blessed relief – the stones, fragmented into small pieces, were passed at last.

A week later, Stephens received her £5,000 reward. Sensibly, she vanished with her mighty haul. Reaction to the news was a mixture of incredulity, curiosity and envy. Books on Stephens and lithotriptic (stone-dissolving) medicines appeared in many languages. Doctors argued over hundreds of cases. Had so many eminent colleagues, as prominent physician Richard Mead claimed, 'acted a part much beneath their character' in falling for 'an old woman's medicine at an exorbitant price'?

The French scientist Sauveur Morand determined that the recipe was essentially alkaline soap and lime – the eggshell and snail-shell powder – rendered less noxious by herbs. He then carefully sawed a bladder stone into four pieces, weighed them and put each piece in a different jar: one filled with normal urine, another with urine from a man taking Stephens's remedy, a third with a soapy solution, and a fourth with the remedy diluted with water.

By modern standards, it was a very small sample. But back then, the very notion of comparative medical trials was exemplary. The jars were kept at roughly body temperature and after a month the stones were weighed again. The stones treated with the remedy had indeed become slightly lighter. But other researchers found the medicine had little effect; still others decided that as lime was the active ingredient, it would be simpler just to drink lime water. These lithotriptic medicines clung on for a while until better treatments emerged. But Stephens became a byword for quackery, and by 1773 the surgeon Percivall Pott could write her off as 'an ignorant, illiberal, drunken, female savage'.

Why the disagreement? Doctors were unwittingly arguing about two different disorders: one for which the Stephens cure was effective and another for which it was not. There are two distinct types of stone: acid stones made of calcium and uric acid and struvite stones ('infection stones') made of magnesium, ammonium and phosphate. These days, stones are often broken up with ultrasound, laser or an endoscope. But modern lithotriptics, such as potassium citrate, are still with us, and their use is telling. Potassium citrate is prescribed to render urine alkaline, so this treatment dissolves acidic stones but not infection stones. The Stephens and lime water cures worked on the same principle. Stephens even recommended that patients refrain from urinating too often, the better to let their newly alkaline bladder contents work on the stones.

In older men, uric acid stones can form when an enlarged prostate interferes with urination and keeps liquid sitting for long periods in the bladder. This may have been what Hartley and the elderly male witnesses at the House of Lords had; the Stephens cure was indeed what such men needed. But depending on which kind of stone you had, the Stephens cure might prove either worthy or worthless, which is why subsequent researchers and patients were so vexed by the conflicting case studies.

So her cure appears real enough, for some people at least. But was it really worth a whopping £5,000? It depends on how long a view you take. The research prompted by Stephens resulted in a most unexpected discovery. In 1752, medical student Joseph Black wrote to his father from the University of Edinburgh about his student thesis, which was to be 'on the Properties & virtues of Lime Water in Dissolving the Stone in the Bladder... as it shows a tendency to remove one of the most excruciating Disorders that render men miserable'. Black had noticed his professors disagreeing with each other over lithotriptic research and had decided new experiments were needed to advance the debate.

First he set about determining what exactly was in lime water. It was thought that lime was made caustic by the addition of a fiery substance, the so-called 'phlogiston'. But, by weighing his sample before and after a reaction, he showed that something had been lost, not gained. That something was a previously unknown gas: carbon dioxide. Black's discovery was a defining moment in the development of chemistry as a modern science. After many twists and turns, the apparently foolhardy investment in an eggshell remedy had fostered a landmark discovery. In the end the canny Mrs Stephens's £5,000 cure was probably worth every penny.

## Hats off to Mr Henley

*In the 1830s, no fashion-conscious woman ventured out of doors without her poke bonnet. Bonnets were big and growing bigger, with crowns built to accommodate increasingly elaborate hairdos and festooned with flowers, feathers and coloured ribbons. But the stand-out feature was the enormous brim, a rigid ruff of cotton or silk sticking up at an angle that defied both gravity and logic. And women weren't alone in their enthusiasm for the wide-brimmed bonnet. Men of science found the bonnet strangely alluring. The springy, silk-covered wire that held those brims aloft was sturdy, flexible and insulated – all ideal qualities for experimenting with electricity. Dock labourer William Henley might seem an even less likely follower of ladies' fashion. But he foresaw a huge demand for insulated wire and when he built a super-efficient version of a bonnet-wire maker's wrapping machine, he paved the way for an industry.*

One day in 1829, William Henley packed a few belongings and climbed aboard the London stagecoach. He was 16 and had had enough of small-town life and his dreary job in the family leatherworks. It was time to try his luck in the metropolis.

As Henley bumped along the rutted road to London, things were happening in the city that would soon change his fortunes. In a laboratory in the basement of the Royal Institution, Michael Faraday was fiddling about with a length of copper wire, some string and a few strips of calico. He was making a copper coil to carry an electric current.

If his experiment was to work, he needed to insulate each turn of the wire from the next. And so, as he shaped the wire, one turn at a time, he wound a cotton string along the sides, slid a piece of calico between the layers and fixed the wrappings in place with varnish. It was a chore.

As the great experimentalist began to get to grips with the principles of current electricity and pondered the relationship between electricity and magnetism, the lad from the leatherworks was labouring in London's docks. Henley could earn half-a-crown a day and usually picked up four days' work each week. On his free day he boned up on physics and chemistry, optics and mechanics and indulged his interest in that most fashionable of phenomena: electricity.

He taught himself how to make his own apparatus and electrostatic machines and in 1830, when an aunt gave him £2, he blew the lot on an old lathe and vice, and a plank to build a workbench. Soon his one-room lodging had become a workshop.

In the grander surroundings of Mayfair, Faraday had also been busy. On 29 August 1831, he discovered electromagnetic induction, the principle behind the electric transformer and the generator. This was the breakthrough that would elevate electricity from scientific curiosity to a powerful technology. But progress would be slow if every piece of electrical wire had to be swaddled in cotton the way Faraday had done it.

According to historian Allan Mills of the University of Leicester, Faraday and his fellow experimenters had made the delicate coils of their earlier galvanometers from ready-insulated wire. They bought this wire from manufacturers of tapestries and other fine furnishings, who had for centuries trimmed their wares with fine copper or silver wires wound about with silk ribbon. Loosely wound, the spiralling ribbon produced a striped effect against the bright metal. To achieve a solid colour, the ribbon was wound more closely. No matter how you bent or shaped the wire, the elastic silk stayed put without exposing the metal core.

It was ideal for a galvanometer, but Faraday needed something more robust when he wound his famous induction ring. He needed a wire that wouldn't blow when he sent larger currents through it, and ideally one made of copper. Silver was a good conductor but too expensive; iron was cheap but a lousy conductor; copper conducted six times as well as iron and was affordable.

The electrical pioneers had to make do with bell wire, a copper wire used to link the elegant bell-pulls of upper-class drawing rooms with the service bells in the servants' quarters. Bell wire was sturdy and flexible but it was bare. And so Faraday had to juggle wire, string and strips of cloth as he tried to insulate his coils. This was a fiddly, time-consuming task and the resulting coil was more than 50 per cent insulation. 'That would be a gross waste of space in solenoids, motors and other emerging electrical instruments,' says Mills.

Faraday and his contemporaries may have been engrossed in their experiments, but not to the exclusion of all else. One thing that hadn't escaped their notice was the changing fashion in hats. And one style in particular caught their eye – the poke bonnet. The bonnet's immense brim was stiffened with an edging of springy iron wire camouflaged by a covering of coloured silk or cotton thread.

Milliners bought in ready-covered wire in the latest colours from the haberdasher, but the people who wrapped the wire for the haberdashers were the bonnet-wire makers. They coated their wire using a simple hand-powered machine that fed the wire through a spindle and wrapped the emerging end in one or two layers of thread. The result was a wire with a thin, flexible coating that wouldn't part when bent. Unfortunately, the wire was iron, but that was easily remedied. 'A lot of people were interested in electricity. It wasn't long before some of them were taking a few pounds of copper wire along to some friendly bonnet-wire maker to wind silk or cotton around it for them,' says Mills.

Some instrument manufacturers ordered more wire than they needed, selling the surplus for a tidy profit. One man, William Ettrick, patented his own version of the bonnet-wire maker's machine, which he claimed would turn out 400 feet of insulated copper wire an hour. But still no one quite grasped how big the electrical industry was going to be – except William Henley.

By 1836 he was a skilled mechanic and had begun to sell some of his instruments. 'He soon realised that this was better than labouring in the docks,' says Mills. And after one satisfied customer, the local chemist, began to display Henley's electrical apparatus in his shop window, orders began to pour in. It was time to give up labouring and go into business.

Henley's instincts were spot on. The market for electrical devices was about to take off, and they all needed insulated copper wire. Soon he had made a wire-wrapping machine with not one set of spools and spindles but six, all turned by a single handle. Even if he sold his wire at half the going rate, he could make £1 a day, four times his docker's wage. 'He saw the commercial possibilities of making large amounts of wire and selling it,' says Mills. 'None of the bonnet-wire makers had acumen enough to see there would be a booming industry to supply. They missed a trick, and when bonnets went out of fashion they went bust.'

The home-made wire-wrapping machine was the start of

what became a vast business in telegraphs and cables. With the machine in operation, Henley hired a man to turn the handle while he made instruments for Charles Wheatstone, who needed an 'intelligent mechanic' to help him develop his electric telegraph system. Within a few years Henley had set up a rival telegraph company, invented a better telegraph machine and run overhead wires the length of the country. By the 1850s, he was king of the cable business, specialising in ocean-spanning submarine cables. He ploughed all his profits into his works until eventually his empire included three cable-laying ships, his own dock, a railway line and an ironworks in Wales.

Unfortunately, it didn't include accountants, managers or investors, and in 1877, after one of his ships had sunk and the ironworks had swallowed up all his money, he went the way of the bonnet-wire makers. But electrical cables never went out of fashion, and although Henley was bankrupt, there were plenty of investors willing to keep his company going. Unlike Henley, they didn't need much foresight to see that demand for cables and wires would just keep on growing.

## ☀ Nothing but a ray of light

*On 27 July 1923 The New York Times's 'Lost & Found' column contained this curious ad: 'FOUR FIELD MICE lost from laboratory, 244 W 74th St., each mouse has a round bald spot on the right side caused by scientific experimentation. $20 reward for each mouse returned dead or alive to Dr. Albert C. Geyser, 244 W 74th St.' New Yorkers were not to know it, but those fugitive mice were harbingers of one of the worst medical disasters of their time.*

Following Wilhelm Roentgen's discovery of X-rays in 1895, doctors around the world turned their primitive X-ray machines on everything from their own hands to patients with cancer and tuberculosis. To Albert Geyser, a brash German immigrant who graduated from a New York medical school in that heady year, X-rays were clearly the future.

Researchers quickly noticed that exposure to X-rays had a remarkable side effect: it made hair fall out. In Austria, physician Leopold Freund recommended it as a treatment for excess body hair, or hypertrichosis. 'Hair begins to fall out in thick tufts when lightly grasped, or it is seen on the towel after the patient's toilet,' he observed in 1899. Unlike painful tweezing and caustic chemicals, Freund pronounced, 'we possess in the Roentgen-treatment an absolutely painless method of epilation'. Tests followed across Europe and North America with apparent success, even 'curing' a 'bearded lady' in Louisville, Kentucky. There were already hints that all was not well, however. In France, some doctors reported that their patients had fallen ill. Loath to admit that X-rays were responsible, Freund blamed 'the hysterical character' of French patients.

Now working at Cornell Medical College in New York, Geyser embraced X-rays with enthusiasm. Like many others, he paid a high price for his zeal: radiologists were belatedly realising that frequent exposure to X-rays could be dangerous, and Geyser suffered burns that claimed the fingers of his left hand. Undeterred, he invented the Cornell tube – an X-ray vacuum tube of leaded glass with a small aperture of common glass, meant to direct lower-energy, or 'ultrasoft', X-rays directly onto a small area of skin. With the Cornell tube, 'the X-ray is robbed of its terrors', declared *The New York Times*. By 1908 Geyser had administered about 5,000 X-ray exposures with his tube, for a variety of skin ailments. Others remained suspicious of X-rays, and the County Medical Society's lawyer warned Geyser that 'the time is coming soon when if a man is burned, the doctor will be held liable… Don't use the X-ray unless you know what you are doing with it.'

Confident that he did know what he was doing with it, Geyser announced in 1915 that he had treated 200 people for hypertrichosis. 'Roentgen therapy is *the* treatment for hypertrichosis,' he insisted in the *Journal for Cutaneous Diseases*, explaining that 'when using the Cornell tube no protection of any kind, either for patient or operator, is needed'. He carried on experimenting, and his doctor son Frank even began offering treatments from his own Manhattan surgery. By 1924, the elder Dr Geyser was ready to formally unveil his hair-removal treatment, and the Tricho Sales Corporation was born.

Ads for the Tricho System were soon everywhere. 'Superfluous hair gone for good,' one proclaimed in the *Oakland Tribune*. 'Newest method… Absolutely painless. No needles.' Hundreds more adverts extolling the virtues of a 'New Electrical Invention' followed in newspapers across North America. 'Artistically reproduces the process of nature… no injury to the skin will result,' promised the *Syracuse Herald*; 'women of refinement' in Erie, Pennsylvania, were told of a 'radio vibration' treatment, and in Canada the women of Winnipeg were promised 'a hair starvation process' so safe that 'Tricho treatments have been given to wives, daughters and sisters of physicians'.

What exactly was this treatment? 'Nothing but a ray of light touches you,' the ads assured readers – though they were curiously vague about just what those rays were.

The Tricho clinics that sprang up in more than 75 American cities gave little away either. Clients sat at a mahogany cabinet with a small front window for the treatment area. The operators, fresh from two weeks' training with Geyser, threw a switch, and then – nothing, save for a faint hum and a whiff of ozone. After a few minutes the machine automatically shut off and the patient booked her next session. Sure enough, their hair fell out. Women were delighted: the New York City clinic alone boasted 20,000 clients. With fees for a course of treatment ranging from a few hundred to over a thousand dollars, the Tricho business was extraordinarily lucrative.

Tricho's triumph was short-lived. The first sign of trouble

was in 1926, when Ida Thomas of Brooklyn sued Frank Geyser for a staggering $100,739 – the cost of her facial treatments plus $100,000 in damages. The reason? After receiving young Frank's prototype treatments in 1920 to 'cure' her facial hair, her skin had inexplicably thickened and wrinkled. The case attracted little attention, but Geyser's son found himself in the news again two years later when he was arrested following a similar complaint. The Geyser family's troubles were beginning to look more serious.

By now doctors were seeing a growing number of women with the same symptoms: wrinkling, mottling, lesions, ulcers and even skin cancer. The signs of X-ray damage were unmistakable. 'In their endeavor to remove a minor blemish, they have incurred a major injury,' concluded the *Journal of the American Medical Association* (*JAMA*). In July 1929 the AMA decided it had seen enough, and formally condemned the Tricho treatment.

Unbowed, Tricho played its trump card: an endorsement by Ann Pennington, glamorous star of that year's hit film *Gold Diggers of Broadway*. And if clients had any lingering doubts, the elder Geyser's impeccable medical credentials probably reassured them. Yet closer inspection of Geyser's record would have shown that, although he carried out research at a prestigious medical college, some of his work was decidedly dubious: he had used electric shocks to treat all sorts of conditions, from gonorrhoea to asthma, and had made unsubstantiated claims to have found cures for tuberculosis and anaemia.

Inevitably, more Tricho victims appeared in *JAMA*, including a patient in Washington DC 'so depressed as a result of the disfigurement of the X-ray burn that she attempted suicide'. Geyser, it seemed, had either been too greedy to heed any warnings, or had convinced himself that his Cornell tubes really were safe. Whatever his motivation, he had installed poorly regulated X-ray machines across the country, and tens of thousands of women – perhaps even more – were exposed to massive doses of radiation on their faces and arms. They had also received wildly varying doses: some women had as few as four treatments, others as many as 50. And because X-ray exposure rises as an inverse square of distance, even a slight shift in sitting position could double or treble a client's dose.

With the prospect of being sued for millions of dollars, the Tricho Sales Corporation collapsed in 1930. But the Tricho story didn't end there.

Emboldened by Tricho's quick profits, copycat operations sprang up in beauty parlours across the US and Canada, bearing such innocent-sounding names as Marton Laboratories, Hair-X and the Dermic Institute. One operator questioned by the authorities in Vancouver could scarcely name a single major technical specification of her machine, let alone who built or serviced it.

Pressure from local medical and business groups drove these operators from view – but not out of business. In 1940, detectives in San Francisco raided what they thought was an illegal abortion clinic. It turned out to be a backstreet hair-removal clinic. Such clinics operated as shadowy cash-only enterprises until at least the 1950s.

Decades later, a second wave of Tricho-related injuries emerged: telltale scarring, wrinkling and cancers that, as one doctor in Toronto put it, were 'obvious stigmata of radiation exposure'. By 1970, American researchers were attributing over one-third of radiation-induced cancers in women to X-ray hair removal.

Given cancer's long latency and the many years that Tricho parlours and their ilk persisted, the procedure may not yet have claimed its final victim. Tricho's most famous customer, though, had reason long ago to regret her endorsement. After spending her final years as a recluse in a small hotel room off Broadway, Ann Pennington died in 1971. Her cause of death, it was reported, was a brain tumour.

# ☀The great tooth robbery

*The night of 18 June 1815 was one to remember. After 23 years of war in Europe, Napoleon faced the combined might of England, Holland and Prussia at Waterloo. By 10 pm, the battle was over. The French were defeated and 50,000 men lay dead or wounded on the battlefield. The casualties were high – but for one group of people that was reason to celebrate. They were the dentists who were about to benefit from the great tooth bonanza.*

Waterloo was a well-timed battle. By the end of the fighting, night was closing in and the battlefield scavengers could go about their work unseen. In the gloom, shadowy figures flitted from corpse to corpse, gathering up the soldiers' weapons and winkling out any valuables tucked inside their torn and bloodied uniforms. Then came the final act of desecration: with expertise many a dental surgeon might envy, they deftly pulled and pocketed any intact front teeth.

Taking teeth from the dead to replace those lost by the living was nothing new. But this time the scale of it was different. The flood of teeth onto the market was so huge that dentures made from second-hand teeth acquired a new name: Waterloo teeth. Far from putting clients off, this was a positive selling point. Better to have teeth from a relatively fit and healthy young man killed by cannonball or sabre than incisors plucked from the jaws of a disease-riddled corpse decaying in the grave or from a hanged man left dangling too long on the gibbet.

Until the 18th century, false teeth were made in much the same way as they had been since the 6th century BC. Then, the most skilled manufacturers of dental prosthetics were the Etruscans. They did a fantastic line in gold bridgework. Depending on the size of the gap, they made a series of gold hoops. The outer ones fitted around the nearest sound teeth, and the rest were filled with artificial teeth carved from ivory or bone and riveted in place with a gold pin. These not only looked impressive, they were secure enough to eat with. The same can't be said of many later designs.

In the late 18th and early 19th centuries people dreaded losing their teeth: the toothless had sunken cheeks and looked old before their time. Without teeth, it was hard to speak intelligibly. In the upper ranks of society, the toothless tended to keep their mouths shut rather than reveal their naked gums. For those who could afford it, the answer was a set of false teeth. But dentures rarely fitted, looked nothing like the real thing and in most cases weren't secure enough to risk eating with. Some sets of teeth were carved from a single piece of ivory or bone. In the more sophisticated designs, artificial teeth were riveted to a plate made of ox bone or hippo ivory. The teeth were carved from the same material – unless dentists could lay their hands on human teeth.

The biggest drawback of all was that the lack of enamel on bone and ivory meant decay soon set in. The result was inevitable: a rotten taste in the mouth and evil-smelling breath. The fashion for fans was prompted by the all-too-common need to hide bad teeth and stinking breath. Dentures made from human teeth were better. They looked better, resisted wear and kept their colour longer – but they were still liable to decay and eventually needed replacing.

What dentists wanted more than anything was a steady supply of human teeth. They could never get enough, so prices were phenomenal. In 1781, Paul Jullion of Gerrard Street in London was charging half a guinea for a single artificial tooth, and four times that for a human one. A row of artificial upper teeth cost £20 and 10 shillings. The real thing fetched an astronomical £31 and 10 shillings.

Sometimes the poor could be persuaded to part with good teeth. In 1783, a dentist advertised in a New York newspaper, offering 2 guineas each for sound teeth. But people had to be desperate to sell their teeth. The dead needed no persuading.

For the discerning patient, teeth from the battlefield were

the best they could hope for. It wasn't always what they got. Many second-hand teeth came from mortuaries, the dissecting room and the gallows. The biggest purveyors of teeth were the 'resurrectionists' who stole corpses to sell to medical schools. Teeth were one of the perks of the job. Even if they dug up a body too far gone for the anatomy classroom, they could still pocket a tidy sum by selling the teeth.

Astley Cooper, the most popular surgeon in London in the early 19th century, kept a whole band of resurrectionists in business. He bought the bodies, but the teeth went elsewhere. According to his nephew Bransby Cooper, author of *The Life of Astley Cooper*, sometimes bodysnatchers didn't even bother to take the body. 'The graves were not always disturbed to obtain possession of the entire body, for the teeth alone at this time offered sufficient remuneration for the trouble and risk incurred in such undertakings. Every dentist in London would at this time purchase teeth from these men.' Needless to say, the dentists would never admit it. Instead, they reassured their patients that their teeth came from the safest of sources – the battlefield.

Before the Battle of Waterloo, the Peninsular War had bolstered supplies. Tooth hunters followed the armies, moving in as soon as the living had left the field. 'Only let there be a battle and there will be no want of teeth; I'll draw them as fast as the men are knocked down,' says one such hunter in *The Life of Astley Cooper*. There were so many spare teeth that they were shipped abroad by the barrel. In 1819, American dentist Levi Spear Parmly, inventor of floss, wrote that he had 'in his possession thousands of teeth extracted from bodies of all ages that have fallen in battle'.

By this time, the first porcelain teeth had begun to appear. Initially they were too white, too brittle and made a horrid grating noise. Then, in 1837, London denture maker Claudius Ash – driven by his hatred of handling dead men's teeth – perfected porcelain dentures and began to manufacture them commercially.

Even so, trade in the real thing continued well into the second half of the century. Supplies increased during the Crimean War of the 1850s. And in 1865 the *Pall Mall Gazette* reported that some London dentists still refused to switch to porcelain. They now had a whole new source: on the other side of the Atlantic the tooth robbers were hard at work, cleaning up behind the armies of the American Civil War.

# 4 Persistence pays

History tells us that some of the greatest scientific insights came in a flash, during a dream, in the bath tub or under an apple tree. But science rarely leaps from one eureka moment to the next. Often the human attribute that serves a scientist best is patience...

## ☀ Lady of longitude

*Mary Edwards sat down at the kitchen table, picked up her quill and began a long series of sums. Every few minutes, she stopped, flicked open a book of tables and began to run her finger down the columns until she reached the figure she needed for the next step in her calculation. Over and over, she added, subtracted and looked up figures until the columns began to wobble and the numbers jiggled on the page. It was time to stop. She couldn't afford to make mistakes.*

For a young country clergyman, John Edwards was beginning to do well for himself. In 1774, he had just been made curate in a little village in the English county of Shropshire. Curates didn't earn that much and Edwards had both a young family and an expensive hobby to support. He was more interested in telescopes than theology, and his attempts to make better instruments were proving costly. No matter. He had ways of supplementing his stipend. He took on paying pupils and he was a paid 'computer' for the *Nautical*

*Almanac*, a collection of data that enabled sailors to plot their ship's course accurately. The first had a drawback: teaching ate into his time with his telescopes. The second suited him better, for as computer historian Mary Croarken discovered, he left most of the work to his wife.

By the 1770s, the *Nautical Almanac* had become an indispensable aid to navigation. The almanac was the brainchild of Nevil Maskelyne, the Astronomer Royal and a member of the Board of Longitude, a body set up in the early years of the 18th century to solve the problem of finding longitude at sea.

Published annually, the almanac contained all the data a sailor needed to pinpoint his ship's position – a sort of ready reckoner for calculating latitude and longitude. The key parts of the almanac were the 'lunars', the predicted distances between the moon and the sun and certain other stars at specified times on a given day in Greenwich – that is, at longitude zero. To a navigator these were as useful as a clock keeping Greenwich time. To find longitude, a navigator needed to know the difference between the time aboard ship – which he could check by the sun – and the time at Greenwich. Every hour of difference is equivalent to 15 degrees longitude. And to find the time at Greenwich, all the navigator had to do was measure the distance between the moon and a suitable star and look the figure up in the almanac.

Before the almanac, voyagers could work out where they were by the lunar distance method, but they needed impressive mathematical skills and detailed knowledge of planetary motion. Even then, the calculations took around 4 hours. Maskelyne's great innovation was to employ a team of human computers to take the slog out of the job by calculating lunar distances at Greenwich in advance. This reduced the shipboard part of the job to a set of simple astronomical measurements and half an hour of easy sums.

The first of Maskelyne's almanacs appeared in 1767. It was a success, and the Board of Longitude wanted more. But it was a lot of work. The daily lunar distances vary from year to year, so a ship leaving on a long voyage might require tables for several years ahead. Maskelyne needed more computers to do such a vast number of calculations. He began to recruit a network of clergymen and schoolmasters, men with a working knowledge of maths and a basic grasp of astronomy. In 1773, Maskelyne was introduced to John Edwards. With his passion for telescopes, and an obvious need to increase his income, he seemed ideal.

The following year, Edwards had a new job as a preacher with a much larger stipend and a house in the Shropshire town of Ludlow. On top of that, he was being paid for computing, on average, six months' worth of lunar tables a year. It was just as well. Edwards was spending more and more time and money on his telescopes. 'He needed the money to buy the metals to make the mirrors for his telescopes,' says Croarken, a visiting fellow at the UK's University of Warwick. Maskelyne might have wondered how Edwards found the time between preaching, teaching and carrying out experiments. It was simple: his wife Mary did the work. 'John's name was on the payroll, but Mary was clearly doing most of the computing from the start,' says Croarken.

So how did an 18th-century clergyman's wife come by these mathematical skills? John may have taught her. But given how quickly and easily Mary slipped into the role of computer, Croarken thinks she must have been good at maths even before she married. Perhaps she had brothers and sat in on their lessons. Perhaps her father was keen on the numerical puzzles and riddles popular at the time, and had maths books around the house.

Mary didn't have to be brilliant at maths, just capable and careful. Maskelyne had done the complicated work in advance, leaving the computers to do little more than long sequences of additions and subtractions. For every entry in a table, Mary might have to look up 12 figures in astronomical tables and perform 14 operations on the data – and then repeat it for every day of the month. 'It was very, very boring work,' says Croarken. But Mary was quick and rarely made mistakes.

In 1784, John's passion for telescopes proved fatal. Experimenting with a new mix of metals to improve the reflectivity of his mirrors, he inhaled a lungful of arsenic fumes and died. Mary and the children now faced poverty and homelessness. John had run up debts. His stipend and the house went to his successor. To Mary, the obvious way to support her family was to carry on computing. But would the Board of Longitude employ a woman?

Mary wrote to Maskelyne. He must have realised that for the past decade Mary had been doing the bulk of the computing. And she was good – perhaps the best of his computers. The work kept on coming.

Struggling to make ends meet, Mary took on more and more work for the almanac, until eventually she was computing a whole year's worth of tables. And while other computers took several months to deliver a two-month chunk of tables, Mary could turn it around in three or four weeks.

Mary's reputation for reliability and accuracy was her greatest asset. When the computers got too far ahead – with 10 years' worth of tables in hand – the Board of Longitude stopped the work. Mary asked for compensation for lost income – and got it. When work started again, Mary was put back on the payroll. In 1811, when Maskelyne died and a new Astronomer Royal was appointed, the steady stream of work slowed to a trickle. Again, she petitioned the board – and again they stepped in to protect her livelihood.

In her researches among the Royal Greenwich Observatory archives, Croarken uncovered more than a woman's struggle to support her family. She also found a Nevil Maskelyne at odds with the character of popular mythology. Maskelyne was a key player in the story of longitude. While humble carpenter John Harrison is portrayed as the hero, famously labouring to create a chronometer that would keep accurate time at sea, Maskelyne has been vilified as an arrogant, elitist astronomer determined to promote the lunar tables at Harrison's expense. But both the chronometer and the tables solved the longitude problem.

'Maskelyne admired Harrison's chronometers,' says Croarken. 'But they were too expensive to be practical.' Until the price came within everyone's reach, lunar tables were a better bet – cheap, reliable and practical. It was for the same reason that Maskelyne employed Mary Edwards. He may have been moved by her predicament. 'But he kept her on because she was good at it, and got the job done,' says Croarken.

## ☀ Fruits of the tomb

*When Giuseppe Passalacqua went to Egypt in the 1820s his plan was to do a bit of horse-trading. He soon discovered a more lucrative line of work, excavating ancient tombs and selling off their contents. While Passalacqua found many priceless treasures, unlike most tomb-robbers he also made off with the more mundane. If something could be carried off, it was – right down to the dried-up offerings left to feed the ancients in the afterlife. Among these were some strange shrivelled fruits that have posed a series of puzzles ever since. They came from some sort of palm tree, but not one anyone recognised. Had the tree vanished along with the pharaohs?*

In 1826 Giuseppe Passalacqua, an Italian horse-trader turned tomb-digger, left Egypt and headed for Paris. His plan was to show off his vast collection of Egyptian antiquities and tempt the French government into buying it for the Louvre. Passalacqua had excavated tombs at several sites in Egypt and had made important discoveries. He was the first to investigate an intact burial, complete with mummy, coffins and funeral offerings, all of which he added to his haul. But although the

French were fascinated by all things Egyptian, they baulked at Passalacqua's price. Disappointed, he took his collection to Berlin, where he sold it to Crown Prince Frederick of Prussia for a knock-down price and a job as director of the Berlin Museum.

Passalacqua's diligence in stripping tombs clean meant there was plenty in his collection for the serious scientist. For Carl Kunth, Berlin's leading botanist of the day, the greatest treasure was the assortment of plant material preserved since the days of the pharaohs. Among the bits and pieces, Kunth was intrigued to find three sorts of palm fruit. He recognised dates and the fruits of the doum palm but he couldn't identify the third. Although he had only dried and shrivelled fruits, Kunth knew they came from a tree that was new to science. He named it *Areca passalacquae*. Others simply called it the Egyptologists' palm.

Eleven years later, German adventurer Prince Paul von Württemberg was exploring the desert of northern Sudan when he discovered a distinctive palm tree bearing masses of deep purple, plum-sized fruits. It would be more than 20 years before botanists connected the prince's tree with Passalacqua's fruits. The Egyptologists' palm is known today as *Medemia argun*, the argun palm.

The palm remained tantalisingly elusive. Occasionally, some doughty explorer would stumble across a few in the Nubian desert of north-east Sudan, one of Africa's driest and most inhospitable places. Two who did, in May 1863, were John Speke and Augustus Grant, fresh from discovering the source of the Nile. Heading back north towards Egypt, they reached the point where the Nile makes a vast westward loop and, bored by what had become a 'tame and monotonous' boat ride, they took a short cut across the desert. Their route led them to a desolate, craggy place not far from the modern border with Egypt, where they were astonished to see a line of unfamiliar palms. The purple fruit with its large seed and thin fleshy covering was inedible, Grant reported, but 'the wood would answer for beams; and we saw our camel-men make shackles for their camels of its leaves, considering them softer for the feet'.

By the end of the 19th century even the sporadic sightings from Sudan had begun to dry up. British colonial officials there warned that the grove found by Speke and Grant was in danger of being destroyed by the local people, who wove matting from the palm leaves. The last specimen sent to the Royal Botanic Gardens at Kew in west London was in 1907. 'Then there were no more,' says Bill Baker, head of palm research at Kew. 'Botanists accepted that it had probably gone extinct.'

One mystery had now given way to another: how had a tree so familiar to the ancient Egyptians vanished so completely? The ancient Egyptians seemed to value it highly. Archaeologists have found the fruits at sites dating from early pharaonic times, around 2500 BC, right up to the 7th century, and stretching all the way along the Nile from the far south of Egypt to the delta. Even King Tutankhamen went to the next world with a supply of argun fruits.

The argun palm, it turned out, had not quite vanished. In 1963, Swedish botanist Vivi Täckholm took a group of students from Cairo University to Dungul oasis, a remote spot in the Egyptian part of the Nubian desert. There they found a single argun palm bearing immense clusters of purplish fruits. A search for more trees revealed only seven small seedlings. The following year, a geologist visiting nearby Nakhila oasis reported a lone tree. For Egyptians, the discoveries had a special significance: a tree that had so long been part of their culture was alive and growing in Egypt. It wasn't extinct – not quite.

Today, the picture has improved slightly. In 1995, two palm-fanciers mounted an expedition to look for the trees that Speke, Grant and other hardy travellers had seen in Sudan. They struck lucky. A local camel-drover knew the tree and where it grew. They found 14 mature trees and 15 seedlings. The following year, the camel-drover took them to a second site with hundreds of argun palms.

These few places where the argun palm survives appear to be the last remnants of the savannah that once covered the Sahara. Around 10,000 years ago, the climate grew drier and the vegetation began to retreat until all that was left were small patches of the most drought-tolerant trees and a few grasses at spots where groundwater comes close to the surface. 'In pharaonic times it was still much greener and there was less desert. The places where the palms are now are what's left of the ancient vegetation,' says Haitham Ibrahim, an ecologist at the Egyptian Environmental Affairs Agency in Aswan. The argun palm probably originated in the region straddling what is now the border between Egypt and Sudan, but was imported and grown throughout Egypt. Why?

Archaeologists think that the way the fruits were offered to the dead suggests they were part of the diet. Grant and Speke reckoned the fruit inedible, but on the expedition to Dungul in 1963 student Loutfy Boulos, now one of Egypt's most eminent botanists, tried them and described them as sweetish and perfectly acceptable, though perhaps not to modern taste. In Sudan desert people still make ropes, matting and baskets from the leaves, which are stronger and more flexible than those of date and doum palms. It's a tradition that probably goes back millennia.

So what are the prospects for Egypt's legendary palm tree? In 1998, a team of Egyptian and German botanists visited Dungul to check on its solitary tree. It was dead. The trunk was still standing but the crown had been blown off. However, the seven original seedlings had matured and there were another 29 small seedlings. Ibrahim has recently discovered a few palms at other places in Egypt. He found not one but two at Nakhila oasis – both male trees – and three sites in wadi Allaqi, east of the Nile, each with just a single tree. Dungul remains the most important refuge for the argun palm in Egypt.

Baker and Ibrahim have made the gruelling trip to Dungul four times to see what might be done to conserve the palms. 'It's not surprising it took so long to find the tree in Egypt,' says Baker. 'You drive south from Aswan for 180 kilometres then turn off the road and drive into the desert for 50 kilometres. The stuff on the ground isn't sand but like fine dust – one wheelspin and you're stuck.'

What's at stake is not just a part of Egypt's cultural heritage but its biodiversity. 'Dungul oasis has fewer than 10 species of flowering plant but that makes it a hotspot of diversity in this bleak landscape,' says Baker. 'In terms of life in the desert, it's hugely important.' Some of Egypt's most endangered animals, such as the extremely rare Nubian ibex and the slender-horned gazelle, may depend on it. 'Dungul is a remarkably lively place. In the morning there are footprints everywhere,' says Baker. The loss of any of the plant species could be catastrophic.

In 2005, the number of palms at Dungul had fallen from 36 to 32. In 2007, only 25 remained. With climate change bringing even more extended dry periods, the tree's future is hanging in the balance. 'We've had a very dry 13 years,' says Ibrahim. 'If we lose a few more trees, what then?'

## ☀ The impossible pound note

*It was about 4 o'clock on 14 November 1888 – a Wednesday. The man behind the counter of the spirit store on Glasgow's Parliamentary Road was pretty sure of the time. He remembered the man too. About 40 years old, medium height with brown whiskers. He had a 'dissipated appearance' and 'appeared to have sore feet'. But it was the brand-new pound note he proffered that aroused the shopkeeper's suspicions. The paper was too soft: when he rubbed it between his fingers and thumb, his fingers went right through it. It had to be bad, he said. The man snatched the note and, despite his sore feet, ran off. This was the second dodgy pound note in two days. The Bank of Scotland was appalled. Three years earlier it had issued a new set of notes incorporating so much chemical cunning that they were impossible to forge. At least, that was the theory.*

All the banks were worried about this new-fangled photography. Banknotes had always been the target of forgers but only the most skilled engravers ever got close to a perfect copy. Photography was going to make the counterfeiter's life a whole lot easier. Take a picture of the note – or its component parts – transfer each negative to a printing stone and print away.

The banks had some defences. They printed in colours that were hard to photograph, such as blue. Some added a series of features in different colours. That wasn't enough for the Bank of Scotland. In 1880, the bank called in Alexander Crum Brown, one of Scotland's most eminent chemists, and asked him to find a foolproof way to foil the forger.

For three-and-a-half years, Crum Brown and his assistant John Gibson experimented in their lab at the University of Edinburgh. First they turned forgers themselves. Notes printed in black on white posed no problem. Blue was harder to photograph – but a yellow filter improved the image. The real challenge was to forge a note from a rival bank, the Royal Bank of Scotland. Its notes were the best-protected of the time and bore a whole set of security devices printed in red, blue and black and all overlying a finely engraved wiggly pattern printed in yellow. But, as Crum Brown proved, even these ingenious notes were not beyond the skills of 'a trained chemist who turned his attention to forgery'.

The trick was to isolate each coloured symbol by removing all others, photograph it and transfer the negative to a printing stone. 'What photography alone cannot do, may be done by photography aided by chemical treatment of the note,' Crum Brown told the bank.

First they needed an image of the wiggly yellow pattern.

Simple. They photographed a small portion of the note that bore nothing but yellow wiggles and printed it repeatedly over a suitable area of paper. Then they dunked the note in a solution that bleached out the yellow, leaving just blue, black and red. To capture an image of the blue feature, they bathed the note in a sequence of chemicals – a dip in dilute acid, followed by the addition of hydrogen sulphide and then a metallic salt. 'Every delicacy of the original device remains in black instead of blue,' said Crum Brown. Black was easy to photograph.

Now they dissolved away the blue to leave just the red and black, and as the two colours didn't overlap Crum Brown covered each in turn and photographed the other. Four negatives on four stones, which could then be printed in the right colours. 'All that was needed to complete the note was the signature of the cashier, and a forger would no doubt add that with a pen,' reported Crum Brown.

At this point, he almost gave up. It seemed there was nothing a forger couldn't crack. Then he hit on an idea. The chink in a banknote's defences was the ink. Because each colour was chemically different, a clever chemist could always find a way to isolate each security feature, by treating the note to remove each colour in turn. But if all the inks were made from the same chemicals, that would be impossible. 'If now we mix our inks so that exactly the same substances are present in them all, but in different proportions, our methods of separation will, as a rule, fail because whatever one ink does they all will do.'

Crum Brown picked three pigments – ultramarine, cadmium yellow and vermilion – which could be mixed to give a range of colours. He also insisted that each coloured feature should overlap to make things even harder for a potential forger.

The bank commissioned a local artist to design a set of notes along the lines Crum Brown proposed. The pound note was brown, yellow and blue and featured the bank's arms, the Arms of Scotland and the Great Seal of Scotland. For added protection, there was a new, improved watermark. When the note was ready, the two chemists did their utmost to forge it. In February 1885 Crum Brown reported back to the bank: 'Our attempts to copy the note in its final form by chemical and photographic processes have completely failed.'

In May the Bank of Scotland issued its new set of notes. They were, as every newspaper in Scotland reported, impossible to forge.

In November 1888, the first dodgy pound notes turned up – one at a bank, one at the Glasgow grog shop, and one at a tea room where a little girl ordered fourpence worth of tea and cakes and went home with 19 shillings and eightpence in good, hard cash. The paper wasn't as crisp or white as it should have been and all the letters were just slightly different. But the forgery would fool most eyes. 'It is a finer piece of work than any forgery they have ever seen at the Bank of England,' said George Waterston, the bank's printer.

The bank warned its staff. The police alerted shopkeepers. A £100 reward was offered for information leading to an arrest. Still more notes turned up, not just in Glasgow but in Leith, Edinburgh, Kirkwall and Dundee. The reward was increased to £250.

In December, the police thought they were onto something. Fake American $20 bills had appeared in Belgium, and the Belgian police were on the trail of an international gang suspected of the crime. One was a dealer in stolen shares. A second had been convicted of passing forged notes in France. This same man had recently ordered a shipment of paper from a London stationer – the sort used to print banknotes.

In February, the Belgian police finally saw one of the dud pound notes and immediately ruled out a connection. The forged $20 bills had been printed by photolithography. The Scottish pound was the work of an engraver. Everyone now agreed that the culprit was a good old-fashioned engraver, not a crook with a camera and a chemistry degree. By the summer, 54 forged notes had turned up.

On 24 July, acting on a tip-off from a Miss Elizabeth Rendall, a draper's assistant in Edinburgh, the police finally found their forger. John Hamilton Mitchell, aged 74, was an artist, engraver and printer. He had recently fallen on hard times and still had three young children to support. At his home, the police found copper plates, a box of paints, an artist's palette and half-finished notes.

Mitchell told the police he had made the notes only to prove he could. 'With the last issue of the Bank of Scotland £1 note there appeared several articles or paragraphs in the newspapers as to the impossibility of forging the notes. Having a natural propensity to overcome anything difficult in my trade, I was induced by that propensity to try.'

Mitchell was jailed for seven years. Miss Rendall collected her £250. Crum Brown's reputation remained intact. His banknotes were still safe from clever counterfeiters with cameras.

## ☀ Land of the midnight sums

*What would a depressed Swedish chemist, alone in the Nordic winter as his marriage to his beautiful research assistant collapsed, do on a Christmas Eve? Go out on the town and find himself a new love? Give way to maudlin sentiment? Well, in Svante Arrhenius's case, no. On 24 December 1894, he rolled up his sleeves and began a marathon of mathematical calculation that took him more than a year, often working 14 hours a day. By the end he had virtually invented the theory of global warming caused by the greenhouse effect, and with it the principles of modern climate modelling. Not only that, his figure for the intensity of warming for a given increase of carbon dioxide in the atmosphere is almost identical to that calculated by the Intergovernmental Panel on Climate Change. Unfortunately, when Arrhenius completed his sums, nobody was remotely interested.*

Svante Arrhenius was an obdurate fellow with a long-standing reputation for rubbing his colleagues up the wrong way. As day-long darkness gave way to months of midnight sun, he laboured on, filling book after book with tedious calculations of the impact of changing concentrations of greenhouse gases on every part of the globe – calculations that would take today's supercomputers only seconds.

'It is unbelievable that so trifling a matter has cost me a full year,' he later confided. But with his wife Sofia gone he had few distractions. And the calculations had become an obsession.

What initially spurred the work was the urge to answer a popular riddle of the day: how the world might have cooled during the ice ages. His calculations showed that a reduction in atmospheric $CO_2$ levels of between a third and a half would cool the planet by 4 to 5 °C, enough to cover most of northern Europe in permanent ice. Eighty years later, researchers found just such levels of $CO_2$ in air trapped in Antarctic ice during the last ice age.

But as he reached the end of his calculations, Arrhenius was also becoming intrigued by the potential of rising concentrations of greenhouse gases to trigger global warming. He concluded that a doubling of atmospheric $CO_2$ would raise world temperatures by an average 5 to 6 °C.

How did he do it? His methods were remarkably close to those used by modern modellers. He started with some basic formulae on the ability of gases such as $CO_2$ and water vapour to trap heat in the atmosphere, derived by 19th-century luminaries such as French mathematician Jean Baptiste Fourier.

The gases, as Fourier knew, let through ultraviolet radiation from the sun. But the Earth's surface converted much of this radiation into infrared heat, which greenhouse gases could trap in the atmosphere, causing it to warm. Fourier never explored this. It was Arrhenius who first realised that the effect was a kind of atmospheric thermostat.

The temperature at which the thermostat was set, he saw, depended on how much ultraviolet radiation the surface of the Earth reflected back into space, and how much it converted into infrared heat. And that depended on whether the surface was coated in ice or ocean, forest or grassland or was shrouded in cloud.

Armed with Fourier's equations, Arrhenius set about dividing the surface of the planet into small squares and assessing the capacity of each segment to absorb and reflect solar radiation. From this he produced a series of temperature predictions for different latitudes and seasons determined by atmospheric concentrations of $CO_2$ that varied from two-thirds to three times the then current value.

Most remarkably, his calculation that a doubling of $CO_2$ levels would cause a warming of around 5 to 6 °C almost exactly mirrors the Intergovernmental Panel on Climate Change's most recent assessment, which puts 5.8 °C at the top of its likely warming range.

Arrhenius published his findings in a paper, 'On the influence of carbonic acid in the air upon the temperature of the ground', presented to the Stockholm Physical Society in 1895. As well as getting the global numbers spectacularly right, he picked out many of the same key elements as modern predictions. High latitudes would experience greater warming than the tropics, he said. Warming would be more marked at night, in winter and over the land.

But he had cracked a problem that seemed to interest no one else. So everyone forgot all about it – until the 1990s, when his century-old work on global warming became one of the most referenced scientific sources of the decade.

Luckily, this labour was but a sideshow in Arrhenius's career. Not long after completing his work on climate, he found fame as the winner of the 1903 Nobel Prize for Chemistry, for work on the electrical conductivity of salt solutions. Soon, too, he had a new wife and child.

Arrhenius was a polymath, dabbling in everything from

immunology to electrical engineering. He was an early investigator of the Northern Lights and a proponent of the idea that the seeds of life could travel through space.

But after the First World War, his mood changed. The optimism of his generation, which had believed that science and technology could solve every problem, crumbled in the face of anxieties about the future. He took up political causes and was an outspoken critic of attempts by both religions and business to stifle the publication of inconvenient scientific results. He campaigned for immunisation against smallpox, and had a passionate interest in agricultural self-sufficiency, fired by the hunger suffered by Swedes during the war.

He railed against the wastefulness of modern industrial society and began to fear that the world would run out of key resources. 'Concern about our raw materials casts a dark shadow over mankind,' he wrote. 'Our descendants surely will censure us for having squandered their just birthright.'

Arrhenius's great fear was that oil supplies would dry up, and he predicted that the USA might pump its last barrel in 1935. He advocated energy efficiency and proposed the development of renewable energy, such as wind and solar power, and engines that ran on alcohol. He sat on a government commission that made Sweden one of the first countries to develop hydroelectric power.

Many Swedes see Arrhenius as a pioneer of modern environmentalism for his efforts to counter the threat of global warming with new forms of energy. But it was not so simple. For one thing, he never made the connection between his work on the greenhouse effect and his later nightmares about disappearing fossil fuels. He knew well enough that burning coal and oil generated greenhouse gases that would build up in the air. But in his earlier writings he had concluded that it would probably take a millennium to cause a significant warming. And when he later began to perceive the scale of industrial exploitation of fossil fuels, he failed to add up the facts: his fear was solely that the resources would run out.

The main reason for his blindness was probably that he thought a spot of warming might be rather a good thing. In 1908 he wrote: 'We may hope to enjoy ages with more equable and better climates, especially as regards the colder regions of the Earth, ages when the Earth will bring forth much more abundant crops for the benefit of a rapidly propagating mankind.'

It is possible that as he wrote this he was harking back to the time he embarked on his mathematical marathon. To someone sitting alone in the bitter cold with an inadequate coal fire, a little heat may well have seemed enticing.

# 5 Fingers crossed

Fortune favours the brave – and there are times when even the best-prepared people find themselves dicing with death. Some know what they are doing, others only realise when it's too late to stop. And for some, the gamble seems worth it whatever the cost.

## ☀ A cure for curates

*Picture the scene. The worthies of the Royal Society gathered around a table. A slightly barmy curate. A sheep, some sharpened quills and a set of silver tubes. It is November 1667, a year after the Great Fire swept through London. In the inner sanctum of English science, a young physician is about to perform a blood transfusion. But what the curate is about to receive is not blood from a fellow human being but from an entirely different species – a sheep. A little blood from a good-natured and docile animal might help to cool the overheated stuff coursing through the curate's veins and so calm his troubled brain. There are a few gasps from the spectators and then it's all over. The curate doesn't seem to have come to any harm. He's having a celebratory drink and a smoke – and says he would be happy to go through it all again.*

Good blood, bad blood, hot blood or cold. If a person's character depends on the nature and quality of their blood, then it ought to be possible to improve it by drawing off some of the old and replacing it with something better.

To 17th-century medics, there was nothing odd about this idea. They felt the same way about blood as their ancient forebears had. A combination of who you were, where you lived and what you ate determined the sort of blood you had – but it could change. Disease corrupted it. Old age made it feeble. Fresh young blood, on the other hand, had the power to rejuvenate. Pharaohs bathed in it to cure them of leprosy. Even Popes believed it had healing powers. In 1492, Pope Innocent VIII was given a draught of blood taken from three young men in a vain attempt to fend off death.

But in 1628, William Harvey discovered that blood circulates around the body in a closed system. People stopped thinking that drinking or bathing in blood would have any effect and began to think about altering the blood itself. In some of the earliest blood-changing experiments, carried out in Oxford in 1656, the young Christopher Wren made a dog drunk with an infusion of wine and beer. It was only a short step from infusion to transfusion, giving blood in place of drugs or tonics.

In February 1665, a young Oxford doctor, Richard Lower, performed the first successful transfusion – from one dog to another. When he repeated his experiment before the Royal Society a few weeks later, he caused quite a stir. 'This did give occasion to many pretty wishes, as of the blood of a Quaker to be let into an Archbishop and such like,' wrote diarist Samuel Pepys. But there were more serious implications. 'It may, if it takes, be of mighty use to man's health,' wrote Pepys, 'for the amending of bad blood by borrowing from a better body.'

In Lower's mind, the better body needn't be another human's; it could just as well belong to an animal. 'For there is no reason to think that the blood of other animals mixes less well with human blood than with animal blood,' said Lower. He planned trials to see the effect of exchanging blood between 'Old and Young, Sick and Healthy, Hot and Cold, Fierce and Fearful, Tame and Wild Animals, and that not only of the same but of differing kinds.'

The English weren't alone in their experiments. The French were close behind. In June 1667, Jean-Baptiste Denis, a mathematician from Montpellier, performed the first human transfusion in Paris before the French equivalent of the Royal Society. The patient, a 15-year-old boy suffering from fever, was given a small amount of blood from a lamb. Like Lower, Denis wasn't worried about mixing blood from different species. In fact, he thought the blood of beasts was healthier because animals were not prone to 'debauchedness and irregularity in eating and drinking'. The boy felt a searing heat along his arm, a sign of an immune reaction to the foreign blood. But he recovered and 'no longer showed that slowness of spirit and heaviness of body' which had troubled him when he was sick.

A few days later, Denis transfused a healthy, middle-aged man who was paid for his services. He too felt a searing heat 'all the way to his armpit', but as soon as the experiment was finished he left to spend his well-earned cash. 'He betook himself to find his comrades and carreyed them to the Tavern,' wrote Denis. 'At noon, finding himself more hearty (whether by the new blood he had received six hours before, or by the quantity of wine he had drank), he fell upon a sort of work so laborious to his whole body that it might almost tire a horse.'

Back in London, Lower decided to try changing a man's character by transfusion – in front of a gathering at the Royal Society. The man was 22-year-old curate Arthur Coga, whose brain was considered 'a little too warm' for a clergyman. Blood from a calm and docile sheep seemed the ideal remedy. Lower drained off a small amount of Coga's blood then connected him up to the sheep with a tube made of quills and silver pipes. Two minutes and half a pint of sheep's blood later it was all over. Coga seemed none the worse for the experience, and returned a few days later to tell the society how he was faring. 'I was pleased to see the person who had his blood taken out,' wrote Pepys. 'He finds himself much better since, and as a new man, but he is cracked a little in the head.'

Lower and his contemporaries had no idea of the risks they were taking. They knew nothing about blood clotting or immunology. They had no reason to suspect the wrong sort of blood could prove fatal. Fortunately, transfusion was banned before the bodies began to pile up.

In December 1667, Denis performed another transfusion – not so much for the benefit of the patient, a manservant called Antoine Mauroy, but for his wife. The couple were newly married, but that didn't stop Mauroy disappearing into the Paris brothels, often for days on end. Denis was asked if he could cure him of his 'madness'. He thought a transfusion of blood from a calf, an animal with a suitably meek temperament, might work. The transfusion went well, and was repeated a few days later. This time, Mauroy felt a violent pain in his arm and his kidneys. His heartbeat became irregular, and he passed bloody urine, a classic reaction to foreign tissue. Mauroy recovered enough to go back to his old ways. His wife insisted on another transfusion. Denis obliged and Mauroy died the next night.

Denis was charged with murder. But during the trial it emerged that Mauroy's wife had poisoned him and Denis was acquitted. Even so, faced with a great outcry over Denis's activities, the court effectively banned transfusion. Britain and Rome followed suit and a great idea had to wait until medical science caught up with it.

# ☀White knuckles in Wyoming

*In the chill of an early August morning, seven men loaded supplies
into a small boat. Then, after arranging for their companions to meet
them downstream, they shoved off into the swirling waters of Wyoming's North Platte river. Warmed by the rising sun, they paddled
easily until they heard a hollow roar echoing from a canyon downstream. The river made a sharp bend, and they were appalled to see the
current slamming squarely into the far wall 'with a great velocity and
so steep a descent, that it had to the eye the appearance of an inclined
plane'. Downstream, the rapids foamed between rocks that would
have spelled doom for a wooden boat. But Lieutenant John Fremont of
the US army hoped to bounce off the rocks without damage.*

If you study a map of America, the name Fremont is hard to
avoid. John C. Fremont, soldier, explorer and map-maker,
gave his name to towns, cities, mountains, rivers and canyons
all over the American west. But Fremont, the 'pathfinder of
the American west', was a pioneer in another way. He was the
first to try river-running in a rubber raft.

The conventional wisdom among river rafters is that
rubber rafts were invented in the late 1930s and became
popular after the Second World War, when people began to
make their own from war surplus inflatable pontoons used to
build floating bridges. But almost a century earlier, Fremont
had signed a $150 purchase order for '1 air army boat or
floater' from Horace Day, a rubber manufacturer in New
Jersey. Along with the boat, Fremont purchased a repair kit
consisting of '2 pieces India rubber cloth' for $19.99 each, plus
'2 pots rubber composition' for 50 cents apiece.

Fremont's primary duty was to survey the country. In
1842, and again in 1843, he and a grouchy German cartogra-
pher called Charles Preuss made thousands of meticulous
measurements that yielded the first mass-produced maps of
the American west. But Fremont was first and foremost an

adventurer. One of his goals was to try out the raft in a series of canyons in present-day Wyoming. There, the North Platte river cuts through a string of granite and sandstone ridges, dropping 300 feet in 15 miles by Fremont's estimate – a gradient that's manageable in a modern boat but inadvisable in a crude and untested raft.

According to Fremont's journal and Day's patent, the raft was made of rubberised linen. At the time, Day was trying to discover the secrets of Charles Goodyear's vulcanisation process for making superior-grade rubber, and while it is doubtful that Day's boat could have held up if it was made of untreated rubber, his patent merely describes the boat as being made of sheets of 'India-rubber fabric'.

The raft was rectangular, 5 feet by 20 feet, with the floor suspended below the flotation chambers. Its four separate air chambers, inflated with a hand bellows, featured tightly sewn seams that had taken two months of work to make airtight. John Allen, a geographer at the University of Wyoming, believes the raft may have had a wooden frame for added rigidity.

Whatever the details of its construction, the boat reeked of chemicals. When Fremont took delivery and triumphantly unpacked it in his parlour, the fumes overcame his pregnant wife Jessie and rendered the house temporarily uninhabitable.

Fremont used the raft twice before venturing into the unscouted canyons of the North Platte. In Kansas he used it as a ferry when he found his way blocked by a wide, rain-swollen river 'with an angry current, yellow and turbid'. Undeterred, he unpacked his boat and began ferrying ox carts across, one at a time. Then, with daylight fading, he decided to double the load for the final crossing.

The raft promptly flipped, dumping crew and equipment into the water. The rescue was tense, but the only major casualties were a quarter of the expedition's sugar and most of its coffee, a loss 'which none but a traveller in a strange and inhospitable country can appreciate', wrote Fremont.

His second attempt was in Wyoming, on the Sweetwater river, a knee-deep trickle that Fremont hoped to float down to its confluence with the North Platte. But the water was too shallow. After dragging the raft over two miles of sandbars, he gave up, repacked it, and trudged on by foot.

At dawn, Fremont sent most of his men overland, planning to meet them for breakfast at the downstream end of the North Platte's first major canyon. In a foolhardy moment of optimism, he packed much of his surveying equipment on the raft, along with irreplaceable botanical specimens and the journals containing his notes.

Soon he was facing the first major rapids, where in places the current was pinched into passages so narrow the boat barely fitted between the rocks. A wooden boat, Fremont declared, would have been dashed to splinters but the raft 'seemed fairly to leap over the falls… We were so delighted with the performance of our boat, and so confident in her powers, that we would not have hesitated to leap a fall of 10 feet with her.'

'We' may have been an overstatement. In one particularly nasty section, Preuss was stumbling along the river bank, trying to keep the expedition's chronometer from getting a soaking. When the narrowing canyon made this impossible, he stuffed the instrument inside his shirt and hoped for the best. Still, Preuss agreed that each successfully negotiated set of rapids produced 'loud jubilation'.

No living rafter has seen Fremont's rapids, which in 1909 disappeared beneath the waters of the Pathfinder Dam. But Preuss described drops of up to 4 feet and water moving 'with great rapidity'. The previous day, Fremont had noticed that the river 'seemed, from its turbid appearance, to be considerably swollen', another danger sign he may not have fully appreciated.

With an experimental boat and a crew of landlubbers, there was only one possible outcome. After a short respite, the river entered a second canyon, worse than the first. Briefly, the joyride resumed. Then the boat struck a big rock. A heartbeat later everyone was in the water. Fremont and one of the crew fetched up in an eddy on one side of the river; everyone else and the boat fetched up on the other.

Fremont had tried to lash down his equipment, but to no avail. 'The current was covered with floating books and boxes, bales of blankets, and scattered articles of clothing,' he wrote. Their precious chronometer, vital for measuring longitude, was ruined. Journals and botanical specimens were missing, and instruments were at the bottom of the river.

The boat, however, had proved that rubber did have advantages by surviving the flip and the subsequent pummelling. Two of Fremont's men retrieved the raft and set off again, trying to rescue as much equipment as possible. But at the next set of rapids, at least one air chamber blew, and the party had to hike out of the canyon. Hours later, safely in camp, Fremont reported that he slept soundly 'after one of the most fatiguing days I have ever experienced'.

Fremont never blamed the boat for the journey's ignominious end. Even after the wreck, he described its performance as 'triumphant'. Preuss didn't blame the boat either. He blamed Fremont for being 'so foolhardy where the terrain was absolutely unknown'.

The following year, Fremont made one more serious attempt at rafting in a new rubber boat. He took it on an easy float downriver to Utah's Great Salt Lake. On the lake, however, Fremont and his companions almost died when they were caught in a storm on a dangerous expanse of open water. As they paddled madly for the shore, the air compartments began to leak badly. In the rush to make the boat, the seams had been glued instead of stitched.

After that, rubber rafts appear to have been used only as ferries. In part, that was because Fremont pushed the technology too far. Rubber rafts need to be firmly inflated for best handling, and the technology of the 1840s simply wasn't up to creating an airtight seam that would withstand the stress, Allen says. But Allen also points out that later explorers were hard-headed engineers conducting railroad surveys. 'They simply weren't into the adventure, as Fremont was. So most people continued to stick with old, tried-and-true methods.'

## ☀ The coachman's knee

*A coachman's lot was not a happy one in 18th-century London. The streets were dirty and crowded. The air was grimy and damp, and the coachman was exposed to the elements day and night. And, as if that wasn't enough, he had a lot of trouble with his legs. Tight gaiters and the constant banging of legs against the edge of the driver's seat weakened the arteries behind the knees, often causing dangerous bulging of the artery wall. If one of these blood-filled balloons burst, the victim bled to death. Why then would the Hunterian Museum at the Royal College of Surgeons in London exhibit a dried and sinewy coachman's leg with such evident pride? As a reminder of the grim life of the working man in squalid Georgian London? A warning to take care of our legs? No, the unprepossessing specimen is a souvenir of a great surgical success story.*

In the late 18th century, the attentions of a surgeon were about as welcome as the services of the hangman. In the days before anaesthetics or antiseptics, patients were quite likely to die of shock. And if they didn't, they often succumbed to infection. Major surgery was something you submitted to only when the alternative was certain death. Doctors – the proper sort, with medical qualifications – had little to do with surgery. That was work for a lower type of person altogether, a knife-wielding butcher rather than a healer.

John Hunter helped to change all that, transforming surgery into a respectable science. His career was remarkable:

at 13 he dropped out of school, but by his forties he was Surgeon Extraordinary to King George III. His patients included the young Lord Byron, who wore a surgical shoe of Hunter's design to disguise his lameness, the composer Haydn and the writer James Boswell, whose drinking played havoc with his liver. But Hunter's reputation as the man who put science into surgery stems from his treatment of some less illustrious patients – the cabbies who drove their carriages through the stinking streets of London.

By the time Hunter worked out how to treat the blood-filled bulges, or aneurysms, in the arteries of the leg, he was a successful surgeon and the most famous teacher of surgery in the country. He had always been an inveterate experi-menter, curious about all aspects of animal anatomy and physiology, and how animals coped with injury and disease. He believed that a surgeon must be able to justify an opera-tion physiologically. If he was going to operate, then he wanted to know what the cause of the problem was and what the effect of his treatment would be. 'Never,' he said, 'perform an operation on another person which, under similar circum-stances, you would not have performed upon yourself.'

At that time, some surgeons tried to treat aneurysms in the leg by tying off the artery immediately above and below the bulge, often opening up the sac and removing the clotting blood inside. The cleaned-up stretch of artery would heal and within a few weeks the ligatures would break away from the vessel – at least, that was the idea. In most cases, the artery burst open within a few hours, and the patient bled to death. 'I have tried it myself more than once… but the event has always been fatal,' said Percivall Pott, one of the eminent surgeons who had taught Hunter. Amputation, with all its attendant horrors, was considered safer.

Hunter suspected that the operation failed in part because the artery was tied too close to the bulging sac. He was convinced that an aneurysm was the result of some underlying disease of the artery and that tying it where the walls were diseased was asking for trouble. The solution, he concluded, was to place a ligature well above the damaged part.

Leading surgeons, including Pott, objected to this idea on the grounds that the limb would be starved of blood and gangrene would set in. Hunter's curiosity about animal physiology, and his fondness for experiments, led him to think otherwise.

One of the puzzles of nature that interested Hunter was how a deer grows new antlers each year. With the King's permission, he tried an experiment on one of the fallow deer bucks in the Royal Park at Richmond. Hunter tied off the main artery supplying the 'velvet', the skin that covers the growing antler, to see what happened. Immediately, the blood stopped pulsing through the artery and the antler grew cold to the touch.

A week later, to Hunter's surprise, the antler was warm again. His ligature around the artery was still in place, but the smaller vessels on either side had expanded and joined together to create a bypass and keep the blood circulating. The fact that the body could compensate for a lost artery by pressing smaller vessels into service was known already, but the speed at which it happened was a revelation. It meant the body could restore the flow of blood fast enough to prevent gangrene, even if an artery was tied off near the top of a limb.

In December 1785, Hunter acquired a new patient, a coachman with a large aneurysm in his leg. The man begged him not to amputate his leg. On the basis of what he had discovered from his experiments, Hunter decided to try a new operation, tying the artery high up in the thigh. Within six months, the man was back in his driver's seat. He died a year later from another occupational disease of coachmen, pneumonia.

In the next two years, Hunter treated three more coachmen. The first died. The others recovered – a near-miraculous cure rate for the time. The fourth man was so ill that at first Hunter refused to operate. Then he changed his mind. The coachman lived another half-century. When he died in 1837, one of Hunter's admirers, Thomas Wormald, saw an opportunity to check out Hunter's handiwork. He asked the coachman's widow if he could remove the leg. She agreed, so it's said, because she had been so grateful for Hunter's treatment of her husband – and so his leg ended up as a prize exhibit in the Hunterian collection, a testament to the surgeon's skills.

Although Hunter performed only a handful of these operations, the technique became a classic and was used routinely for more than a century. Today's surgeons have more options. With modern anaesthetics and drugs to combat infection, they have plenty of time to repair or replace damaged vessels with grafts. But for patients who are too frail for major surgery, or who have a fast-growing, potentially leaking, aneurysm, Hunter's operation is still the best chance of saving their limb.

# Into the mouth of hell

*After months of preparation, Friar Blas del Castillo could put it off no longer. On 13 April 1538, he tied his habit tightly round himself, crossed his stole over his chest and jammed a metal helmet on his head. Equipped with only a hammer, a flask of wine and a wooden cross, he climbed into the waiting basket and prayed as his three companions lowered him into the crater of one of Nicaragua's most active volcanoes. The local people believed the volcano was a goddess; the Spanish conquistadores thought it was the gateway to hell. But the intrepid monk was convinced the strange lake at the bottom of the crater was filled with liquid gold.*

Friar Blas del Castillo knew that going into the crater of an active volcano was dangerous. He didn't know much about volcanoes and nothing at all about lava lakes, but he had looked over the crater's edge. He had seen the red glow, the smoking fissures and the spitting, bubbling lake of fire. He had smelled the sulphurous gases. And he had heard the stories.

Spanish forces had marched into Nicaragua in 1522. Two years later, the short, squat mountain they called Masaya erupted. The volcano, one of the strangest and most active in Nicaragua, had been the subject of both fear and fascination ever since. It had twin craters, one of which contained a permanent lava lake. At night, the mountain lit up as incandescent smoke poured from the summit. Gases came hissing from cracks in the mountainside. Those who dared to peer over the crater's edge saw red-hot lava, sometimes heaving gently, sometimes boiling, and every so often erupting into fiery fountains.

The Spaniards learned that the Indians thought the volcano was a goddess and made human sacrifices to her. Local chiefs, it was said, sought advice from an ugly old sorceress who lived in the volcano. Unfamiliar with volcanoes, and convinced the sorceress was the devil, the conquistadores concluded that the fiery lake was the mouth of hell.

So pervasive was the belief, that in 1529 Friar Francisco de Bobadilla planted a cross on the mountain overlooking the lava lake to exorcise the devil. That same year, Gonzalo Fernández de Oviedo y Valdéz, one of Spain's most famous chroniclers, climbed the mountain 'to see the fire'. At the top he found an abyss 'so big and round that no shotgun would in my opinion reach from one side to the other'. On the crater's south side was a pit where the floor had collapsed. 'At the bottom of that hole there was a fire that was liquid as water and that matter was burning more fiercely than red-hot coal and more ardent in colour; burning more than any fire can ever burn, if that is possible.'

Blas knew all this, but he had also heard rumours that the pool contained gold. He had looked into the crater and seen bright crusts that looked like melted silver, and a seam of yellow that shone like gold. For such riches, he was willing to risk the heat and the gas – and perhaps even a brush with the devil.

The friar studied the crater for months, then secretly ferried up the gear he would need for a trip into the crater to collect samples. On 13 April 1538, three men lowered him down to the crater floor in a basket. When he reached the bottom, he climbed out, kissed the cindery ground and set about hammering lumps off the shiny crusts and yellow seam. The lake, he decided, was too dangerous to sample without help, and after three hours in the crater he asked his men to haul him out.

Three days later, the friar and his men were back. This time they all clambered into the crater, and rigged up a system of pulleys that would allow them to lower a cauldron into the lake. When the pot touched the lava, it stuck. With an enormous effort, the men pulled it free, retrieving a small amount of molten lava and a few glowing cinders. A third expedition with four more men produced a few extra lumps of half-melted lava before the cauldron stuck fast and the chain broke. Analysis of the samples found only worthless black stone.

On the face of it, the story of Blas del Castillo and his search for gold is a simple parable of greed. For José Viramonte, a vulcanologist at Salta National University in Argentina, and Jaime Incer-Barquero, geographer, historian and former Nicaraguan environment minister, the friar's report of what he saw contains a wealth of useful detail about the volcano's past activity.

Masaya is one of a string of volcanoes that stretches from Guatemala in the north to Costa Rica in the south. This particular volcano is unusual in that it consists of a nested series of cones and craters clustered in the middle of a vast caldera, itself created by a catastrophic eruption thousands of years ago. In the 16th century, there were two active craters, one confusingly called Masaya, the other Nindiri. Today, both are extinct and the volcano's activity centres on a newer crater called Santiago. Vulcanologists can infer a volcano's eruptive history by mapping and dating the many lava flows and explosive deposits. But it's hard to beat eyewitness accounts with precise dates, and the early colonists' fascination with the volcano meant they left plenty of them. Viramonte and Incer-Barquero have translated some of these accounts from the original Old Spanish, and found that, leaving aside discussions of deities and devils, they are impressively detailed and accurate.

It's clear from Oviedo's account that the lava lake was in the Nindiri crater. Some observers recorded the changing shape and size of the vent, while others measured the depth of the crater and the size and position of the lava lake. From these details it's apparent that the lake wasn't always in the same part of the crater and that its level varied enormously. In 1670, lava rose to the top of the crater and overflowed, at which point the bottom of the crater collapsed and the lake vanished altogether.

The last recorded eruption from the Masaya crater was in 1772, when lava flowed non-stop for eight days. Famously, the local bishop headed up the mountain bearing an image of the Christ of Nindiri and ordered the lava to stop. It did. In 1853, after another collapse, the Santiago crater began to form. It has been extremely active ever since.

So how dangerous were the friar's trips into the 'Mouth of Hell'? Ken Sims, a vulcanologist at Woods Hole Oceanographic Institution in Massachusetts, has a good idea. He travels the world collecting samples of gas from volcanoes, and has climbed into the Santiago crater five times. Although it has no permanent lava lake, it has much in common with the Nindiri crater of the 16th century: it's hot, constantly belches out choking gases, and could spew out hot rock and ash at any time. Unlike the monk and his men, Sims is an experienced climber, has state-of-the-art equipment and only enters volcanoes that are constantly monitored for signs of activity.

The Nindiri crater was probably never so deep, nor its walls so sheer, suggests Sims, but the crater's edge would have been just as unstable and prone to collapses. Once on the crater floor, Blas would have been in danger from falling rocks and gas, and there was always the risk of being hit by a fountain of molten lava. 'Messing about with a lava lake – that's impressive,' says Sims. 'A lava lake degasses phenomenally and the fumes are horrible. They burn your eyes and if you aren't wearing a mask can cause respiratory problems.'

What did he think of the friar's sampling methods? 'Dunking a bucket in a pool of lava? I'd say it was insane,' says Sims. But then he admits he has tried it himself. 'During an expedition to Mount Etna a colleague wanted some lava samples but we had no protective fireman suits or probes – so we went off to a hardware store and bought a bucket and some chain. I got the job of throwing the bucket in.' When the lava is viscous, as it is at Etna and Masaya, it's like fishing for boulders embedded in glue. 'Our bucket just bounced off the surface. We tried a few times then gave up.'

After reading Blas's original account of his adventure, Viramonte has nothing but admiration for the man who unwittingly became America's first vulcanologist: 'Given the technology of the time, what he did was a real feat – on a par with landing men on the moon in the 20th century.'

## ☀ Arsonist by appointment

*London has had not one Great Fire, but many. The fire of 1212 exacted an appalling death toll of 3,000; the famous conflagration of 1666 killed only six. For this dramatic reduction in mortality, Londoners could thank centuries of increasingly strict building codes – which, if they did not save most of the city, at least slowed fires enough to give people time to escape. But architects still grappled with the problem of fire. The best solution in the mid-18th century, the vaulted masonry ceiling, was limited by its expense to the most important public and commercial buildings. Fireproofing for homes and businesses was still desperately needed. In 1774, in an attempt to prove he had the answer, one member of parliament carried out a novel experiment: he set his house on fire – with the king and queen inside it.*

David Hartley is forgotten now, but both the USA and Britain owe him a debt of gratitude, albeit for completely different reasons. The son of an eminent philosopher of the same name, the younger Hartley inherited his father's eccentricities and keen scientific mind. He also cut a distinctive figure on the street. Hartley refused to powder his hair which, as one friend commented, made him 'a perfect phenomena at the time'. And he insisted on wearing stockings with the feet cut out, a practice he declared was conducive to good health and 'favourable to pedestrian exercise'.

In 1774, Hartley was elected an MP for Hull. He was notoriously long-winded: biographer George Guttridge noted that 'Lord Liverpool left the House during a speech by Hartley, and after visiting his residence out of town, dined, and returned some five hours later to find the speaker in the same attitude addressing a grievously depleted audience'.

Hartley was respected for his deeply held beliefs, however. He was the first MP to go on record with a bill opposing slavery, decades before others took up the cause, and he was a passionate critic of Britain's policies in America. 'You gave them no alternative but independence or unconditional submission,' he berated colleagues in the House of Commons in 1777. The Treaty of Paris ending the War of American Independence in 1783 was negotiated in part by Hartley and signed in his hotel room, for he was seen as the most likely person to be trusted by the American delegation.

But Hartley's greatest impact in London began before he

entered Parliament. In 1773, he was granted a patent for 'Securing Buildings and Ships From Fire'. Focusing on the vulnerable joists under the floorboards, he proposed sheathing them in thin layers of iron plating. 'A quantity of dry Rubbish of any Kind [such as sand or chalk] should be put over the Ceiling Plates,' he added, 'which will deaden sound between Floors, and at the same Time, still further stop the Progress of Fire.' Fireproofing the joists would help prevent the floor from collapsing, and this combined with the dry filler would be effective in 'stopping the free Supply and Current of Air, without which, no Fire can get to any great Height, or make any destructive Progress'.

The plates themselves were elegantly simple: thin iron plates overlapped the top of the joists, and were held firmly in place by nails pounded through the floorboards and into these joists. Hartley's fireproofing was also cheap. He estimated it would add 4 per cent to the cost of a building, and could be retrofitted to existing buildings, whether they were 'a Magazine, a Merchant's Warehouse, a Banker's Shop, or a private Dwelling-house'.

The benefits of his invention, he argued, would be immeasurable. 'A single Fire-Plate under a Crevice in a Floor, or over a Crack in a Ceiling, might have prevented the Fire of London,' Hartley claimed in a pamphlet in 1774. This made his fellow MPs sit up and take notice. What Hartley was proposing could save entire cities. Parliament voted him a grant of £2,500 to continue his experiments, and passed a special act to extend his patent from 15 to 31 years.

The best way to refine and prove his fireproofing, Hartley decided, was by setting houses on fire. He used his grant to erect and furnish two fire-plated houses in the Berkshire town of Buckleberry for the seemingly perverse purpose of burning them down. And so it was that one Saturday in April 1775 an entirely preventable fire broke out in Buckleberry. Hartley torched the furniture and wainscoting in one room of his house until, as one newspaper reported, 'the inside of the room appeared a perfect mass of Fire'. Hartley invited witnesses inside his house as the fire burnt itself out inside the test room, leaving the rest of the house and its visitors quite untouched. A few weeks later, Hartley repeated his test with the same results.

Emboldened by his trials in the countryside, Hartley erected another fire-plated house where no one could miss it. 'I have built an House upon Wimbledon Common that every one may see repeated Trials and Proofs,' he announced in 1776. The Fireproof House attracted crowds of admirers, including the king and queen. George III was so delighted with Hartley's results that, visiting Wimbledon one morning, he decided not just to see the house for himself but to test it out – with the royal family inside.

'Their majesties, with the Princes and Princesses, first breakfasted in one of the rooms,' reported a witness. 'The tea kettle was boiled upon a fire made on the floor of an opposite room, which apartment they afterwards entered and saw a bed set on fire, the curtains of which were consumed with part of the bedstead... Their majesties then went downstairs and saw a horse-shoe forged in a fire made upon the floor, as also a large faggot was hung on the ceiling instead of a curtain; after this two fires were made upon the staircase and one under the stairs, all of which burnt out quietly without spreading.'

The royal family repaired to an upstairs room, while Hartley filled the downstairs with pitch, tar and kindling, and quickly had a roaring fire going. This soon went out, and the king and his family emerged none the worse for their experience. Hartley's experiment was a resounding success, and fire plates were subsequently used throughout Britain.

As improved fireproofing measures such as cast-iron columns and terracotta roof tiles appeared in the 19th century, Hartley's Fire Plates fell into disuse. 'They may have performed OK under certain conditions – as in, better than nothing – but generally thin iron didn't work as intended,' says Sara Wermeil, an expert on the history of fireproofing.

Hartley faded from public view after the Treaty of Paris. He lived with his sister, tinkered with inventions and hid from would-be visitors. Eventually, the Fireproof House came to a predictably ironic end. The builders of Wildcroft Manor, a mansion erected around Hartley's old test house, did not have the prudence to install fire plates. On the morning of 21 December 1791 Hartley rushed out to Wimbledon Common to find the mansion ablaze. Soon it collapsed onto the fireproofed section, which, with the inrush of air... caught fire.

# 6 Don't call us ...

If the course of true love has never run smooth, neither has the course of science over the centuries. For every brilliant idea and world-changing breakthrough there are hundreds that fell by the wayside. Some were doomed from the start. Others really did seem a good idea at the time.

## ☀ Farmer Buckley's exploding trousers

*In 1931, the peace and quiet of the New Zealand countryside was shattered by a terrifying new phenomenon: suddenly and apparently at random, men's trousers began to explode. Some pairs detonated on the washing line, others as they dried in front of the fire. Worse, some were occupied when they started to smoulder. At first there were just a few isolated reports, but soon the nation was in the grip of an epidemic of exploding trousers. The self-destructive garments all belonged to farmers who had been trying to clear ragwort from their pastures. This pernicious weed had reached New Zealand decades earlier and was now running riot across the country. The latest weapon in the war on the weed was sodium chlorate. But when combined with organic material, such as cotton and woollen fibres, the mixture becomes violently explosive.*

Richard Buckley was lucky. When his trousers blew up he wasn't wearing them. He was badly shocked, but as the *Hawera Star* reported on 12 August 1931, his quick thinking saved him from serious injury. 'While Mr Richard Buckley's trousers were drying before the fire recently, they exploded with a loud report. Although partially stunned by the force of the explosion, he had sufficient presence of mind to seize the garments and hurl them from the house, where they smouldered on the lawn with a series of minor detonations.'

According to the *Star*, there was only one suspect in the case: sodium chlorate. Until recently the chemical had been more familiar to quarrymen than farmers, but when government scientists declared it the best ragwort killer they had seen, farmers began clamouring for it. Buckley, who farmed in Taranaki, on the western side of the North Island, was just one of a growing number of people falling foul of the new, miracle weedkiller. And not everyone got off so lightly. Some were injured and a few died. In one tragic case, a farm worker who wanted to check on his sleeping baby struck a match to see by. His clothes went up in flames and he died a few days later.

According to New Zealand historian James Watson of Massey University in Palmerston North, there was more to the rash of explosions than dodgy weedkiller and bad advice from the authorities. Ragwort, *Senecio jacobaea*, was accidentally introduced to New Zealand in the late 19th century and, like so many foreign invaders, quickly became a pest. By the 1920s, the weed was rampant. What made matters worse was that its proliferation coincided with sweeping changes in agriculture. 'There was a massive shift from sheep farming to dairying,' says Watson, 'and that meant ragwort was an even bigger problem.'

Ragwort contains a battery of noxious alkaloids: it is so toxic that even honey made from its flowers is poisonous. Livestock usually avoid the plant, but once it displaces grass and clover the animals have little else to eat. Sheep can eat it for months before showing signs of illness, but cattle and horses sicken quickly and can die of liver failure.

The boom in dairy farming followed the arrival of new

technology – first refrigerated ships, then motor vehicles and machines to separate cream from milk. The first refrigerated ships began to carry produce to the UK in 1882. At first they took mainly meat. Shipments of dairy produce only took off once motor vehicles began to replace horses, allowing farmers to get fresh milk to the local dairy factory. When farmers began to separate the cream themselves, the butter factories introduced collection rounds, picking up the cream from the farm gate. As demand and factories grew, dairy farms proliferated and spread into remote areas once thought too marginal to bother with. Between 1899 and 1919 the number of dairy cows doubled. Over the next two decades it doubled again.

In the past, farmers had grubbed up ragwort by hand, a labour-intensive job that brought only temporary relief: any roots left in the soil simply re-sprouted. But hands were becoming hard to find. Even the unemployed baulked at the hard work and poor pay. By the late 1920s, farmers couldn't even turn to their families. Farmers who put their wives to work in the fields were frowned on. It wasn't respectable. Nor could they rely on their children. School had become compulsory and there were buses to collect children from the farms and inspectors with cars to check on any who didn't turn up.

In any case, the one-man operation was part of the New Zealand farmer's ethos. They adopted the latest labour-saving devices and kept abreast of agricultural research. And they were prepared to consider any method of getting rid of ragwort. 'When sodium chlorate came along they saw it as a means of rescue. They would try anything that would save them from hiring labour,' explains Watson.

The first the farmers heard of the weedkiller was in 1930, when a scientist at the Department of Agriculture wrote an article extolling its virtues. When a government expert said the weedkiller was far superior at killing ragwort than any other and would, 'where properly applied... completely destroy all the plants', then farmers listened. Within a year, imports soared from almost nothing to hundreds of tonnes.

The accidents started immediately. Mixed with organic material such as the fibres of a farmer's working clothes, sodium chlorate is extremely dangerous, forming compounds that will detonate at the first sign of a spark or a glowing cigarette. Sometimes just a shock or a knock is enough. Washing contaminated clothes made little difference.

Why had the government recommended such a dangerous substance? 'The scientists were more concerned with how effective it was against ragwort than how dangerous it was,' says Watson. This was a time when arsenic was recommended for getting rid of another serious weed, the blackberry. New Zealand's farmers were prepared to take the risks, and they didn't blame the authorities. 'Quite the opposite. They'd have been annoyed if the government had got in the way of them using it,' says Watson. 'If they hadn't had it, many would have been forced off the land. The accidents were the price they paid to keep their farms in production.'

Besides, by the time Buckley's trousers hit the headlines in Hawera, farmers must have known of the dangers of sodium chlorate. 'Most men smoked then and that would undoubtedly have brought it to their attention,' says Watson. 'You can't deal with it for long without discovering that it's dangerous.'

Farmers continued to use the weedkiller until the late 1930s, when word got around that those same government scientists were arguing about just how effective sodium chlorate really was. 'Once farmers heard that it wasn't all that good, they started looking for alternatives.' They are still looking.

Ragwort remains a serious pest in New Zealand. Today's weapons of choice are not herbicides but insects, the plant's natural enemies in its native Europe. The ragwort flea beetle was introduced in the early 1980s and has been very successful in some parts of the country. But the beetles didn't take to the wetter climate of western New Zealand. In 2006, scientists began to release two European species of moth better suited to a wetter climate. The caterpillars of plume moths attack ragwort's roots. Plants attacked by the crown-boring moth produce fewer seeds and their growth is restricted. Insects may prove the much-needed miracle cure – but this time round, scientists are keeping a more careful watch and a tighter rein on them.

## ☀ Like a lead balloon

*In 1844, Parisians with an extra franc in their pocket could wander to the outskirts of the city and buy entry to a mysterious building on Impasse du Maine, a narrow dead-end street just behind the new railway station at Montparnasse. Inside a cavernous hangar, proprietor Edmond Marey-Monge and his team of workmen laboured away, soldering together long sheets of metal to make a giant sphere. Just what this contraption was for only became clear after examining the blueprints: valves for introducing hydrogen and attachments for a passenger gondola hinted at a new mode of transport. The gleaming sphere, Marey-Monge announced, was a brass balloon.*

Metal airships are one of the oldest notions in aeronautics. As early as 1670, Italian mathematician Francesco Lana published his *Demonstration of the Feasibility of Constructing a Ship With Rudder and Sails, Which Will Sail Through the Air*. Lana proposed evacuating the air from a set of copper spheres, which he reasoned would weigh less than the surrounding air and would ascend until the weight of the sphere reached equilibrium with the surrounding atmosphere. He calculated that four vacuum spheres, each with a diameter of 7.5 metres, could lift a boat carrying six passengers.

Lana, alas, could not procure the copper spheres himself because of his obligations as a Jesuit. 'I would have willingly [built it] before publishing these my inventions,' he explained, 'had not my vows of poverty prevented my expending 100 ducats.'

Lana's notion was not entirely fanciful. Recent experiments had demonstrated that air had weight, and in 1650 the German inventor and scientist Otto von Guericke had drawn together small copper hemispheres with such a strong vacuum that teams of horses could not pull them apart. Still, there was a difference between creating a sturdy 50-centimetre sphere and fashioning 7.5-metre globes that could rise into the clouds. As Lana's compatriot Giovanni Borelli pointed out, for Lana's spheres to be light enough to fly, the copper would have to be so thin that on evacuation they would be crushed by atmospheric pressure.

Nonetheless, Lana's idea transfixed Europe for the next century. By 1672, translations and engravings depicting his balloon had appeared in Germany, and in London Robert Hooke penned an English translation. A century later, Lana's copper balloons were still celebrated in poems. They appear in a German interplanetary travelogue of 1744 envisaging trips to Mars, and in 1768 an epic poem in Latin predicted their use in airlifts to rescue those 'shaken by repeated earthquakes'. But with the first successful flight of a hot-air balloon in Paris in 1783, Lana's concept of a rigid vacuum airship fell by the wayside.

Except, that is, in Paris itself. In 1844, inventor Edmond Marey-Monge suggested in the journal *Comptes Rendus* that for 'aerial navigation to be able to render the same services as the navy', balloons must be in 'a position to resist, like our ships, the bad weather of ten or fifteen years of service'. Unlike fragile balloons made of fabric, metal airships could stay aloft indefinitely while performing their commercial or military duties.

After three years working in a custom-built hangar at number 10 Impasse du Maine, on what was then the southern edge of Paris, Marey-Monge was ready to fulfil Lana's vision.

Rather than extract the air from a set of spheres, he planned to fill a single giant sphere with hydrogen. His balloon would be 10 metres across and made from 0.1-millimetre-thick brass sheets banded and soldered together. 'I prefer the spherical form,' Marey-Monge explained, 'because under the smallest surface it contains the greatest capacity', and so, kilo for kilo of brass, gets the greatest lift from the gas it contains.

And lift was just what Marey-Monge needed, because despite the thinness of the metal, his balloon weighed 400 kilograms. His choice of material had several other drawbacks: not only were the curved brass plates difficult to solder together, but French foundries were unable to create the long, thin panels he needed. Marey-Monge had to import brass sheets, each 5 metres long and 50 centimetres wide, from a Prussian foundry. Then he had to construct a temporary wooden frame for workers to build the sphere around. To cap it all, the sheets were so thin, they developed countless tiny holes, and to prevent leaks he was forced to line the interior with thin layers of tissue, glue and varnish, adding another 16 kilograms to its weight.

Marey-Monge planned to launch his balloon on 2 June 1844. Extracting the wooden supports from inside the globe proved a delicate operation, as the balloon's skin was thin and easily damaged. But at last the vast globe seemed ready to fly. And so, with a turn of the valve, Marey-Monge filled his contraption with the 523 cubic metres of hydrogen it would need to break free of the Earth. And then... nothing. It wasn't budging.

Marey-Monge thrust his head through the hole at the base of the balloon and into the dark chamber. The air at the bottom of the sphere was still perfectly breathable, so much so that he could stand there for a full 15 minutes contemplating the cruelties of gravity. Pumping in the precious remainder of his hydrogen supply did not help: when he thrust his head inside again he could hear the hiss of lost money as the gas escaped through leaks in the brass skin.

The balloon never did fly, and Marey-Monge wound up selling the brass for scrap. It was little consolation for a project that had taken three years and cost him a fortune. Despite its ignominious end, the brass balloon had fired the Victorian imagination. Patents for metal airships appeared with some regularity, including a proposal in 1887 for a steel blimp 126 metres long. Yet even Marey-Monge eventually concluded that metal airships would remain unworkably fragile and leaky until a metal light enough to allow overlapping layers could be procured.

By 1897 a candidate had presented itself. Amid Germany's pioneering work in Zeppelin construction, designer David Schwarz conceived of a daring approach: why not an aluminium dirigible? After building two prototypes, with the help of the Prussian Airship Battalion he launched a 38-metre airship from Tempelhof Field in Berlin. Built in the familiar 'blimp' shape, it contained bags of hydrogen within a skin of riveted aluminium plates 0.2 millimetres thick.

Incredibly, it worked. On 3 November 1897, Schwarz's airship lumbered aloft, and sailed 6 kilometres into the German countryside. But mechanical mishaps and its rigid design meant that instead of gently nudging to a stop, the metal colossus crumpled into a field like so much expensive foil.

Despite this, rigidity is one of the great attractions of metal-clads. Because their skin is less prone to deformation at higher speeds, a sleek metal-clad should be able to fly much faster than a textile-covered blimp. Goodyear's original airship designer, Ralph Upson, certainly thought so. In 1929, he formed the Metalclad Airship Corporation to build an aluminium-clad helium airship, the ZMC-2, for the US navy. Although notoriously difficult to handle, the 'Tin Bubble', as it was dubbed, could reach a speed of 100 kilometres an hour, and it put in 2,200 flight hours before it was decommissioned in 1941.

Although metal-clads were too expensive to build and too

difficult to handle, the dream never quite died. Upson's engineer Vladimir Pavlecka continued to patent innovations in the hope of a relaunch right until his death in 1980. Other designs have included one patented in 1964 that promised 'jet-propelled dirigible airships'. Such notions are still floated wistfully by airship designers at conferences.

Curiously, one man who remained ultimately unconvinced was none other than Francesco Lana himself, who feared what his metal airship might be capable of. 'Fortresses and cities could thus be destroyed,' he wrote, 'as iron weights, fireballs and bombs could be hurled from a great height.' So while airships were conceivable, the priest decided in 1670 that the havoc they would wreak meant that humans would never leave the ground. 'God,' he concluded, 'would surely never allow such a machine to be successful.'

## ☀ Night of the mosquito hunters

*If there are 250,000 bats in a colony and each one eats 3,375 mosquitoes in a night, how long will it take the bats to rid one Texan swamp of malaria? Charles Campbell, a doctor in San Antonio, did the sums and was convinced that bats could consume enough mosquitoes to eradicate the disease from the swamplands around the city. But how could he persuade some of the millions of bats that flew in from Mexico each spring to settle where the mosquitoes were worst? Campbell's answer was the bat tower, a high-rise home to tempt even the most discerning bat. And not only would the tower's tenants transform the malarial mires into wholesome and habitable land, they would provide a steady income. Bat guano was a highly prized fertiliser, and if one bat produced 40 grams of guano a year, then a quarter of a million bats …*

It seemed so simple. Mosquitoes spread malaria. Bats eat mosquitoes. Set the bats on the mosquitoes and the disease would disappear. Charles Campbell knew about malaria: he was a doctor. He also knew about bats, which he studied in his spare time. There was no doubt that they ate mosquitoes, and lots of them. He had poked through countless pellets of bat guano and totted up the fragments of wings, legs and other indigestible bits to come up with an estimated nightly toll.

At the turn of the 20th century, malaria was still rife in America's southern states. Each year, there were hundreds of thousands of cases, costing the nation around $250 million. But while parts of Texas were plagued by mosquitoes, the state was also home to huge populations of bats. Late each February, a hundred million Mexican free-tailed bats fly north from Mexico to central Texas, forming immense colonies in caves, old barns and derelict buildings. Campbell's plan was to install bats in purpose-built roosts close to places where mosquitoes bred.

In 1902, Campbell lined some wooden boxes with guano-coated cheesecloth and fixed them in old buildings, under country bridges, and in trees near a cave where millions of bats roosted. After five years and no bats, Campbell conceded that the boxes were a dismal failure. Bats, he deduced, preferred larger homes.

Within months he had built a 10-metre tower at the local experimental farm. It cost him $500. Inside were sloping shelves for the bats to cling to, a large heap of guano to make them feel at home and 'three perfectly good hams with a nice slice cut out of the side of each, exhibiting their splendid quality for the delectation of the intended guests'. But there were no guests. Eventually Campbell resorted to kidnap, capturing 500 bats and incarcerating them in the tower, hoping their squeaks would attract others. By 1910, it was clear that bats were never going to move in. He dismantled the tower and sold the timber for $45.

Disappointed, Campbell left his practice and headed for the wilds of Texas to learn more about bats. He discovered that they don't hunt immediately outside their roost but make a fast getaway to avoid predators that might be lying in wait for them. In April 1911, Campbell built his second tower on the shore of Mitchell's Lake, just south of San Antonio.

'No swamp in the low lands could possibly be worse,' he said. All San Antonio's sewage flowed into the lake, and water seeping from it formed a huge shallow pool – a perfect breeding place for mosquitoes, and the right sort of distance from the tower. In summer great clouds of mosquitoes drove the tenant farmers from their fields around the lake and tormented their animals. Crops went to ruin. Cows stopped producing milk and hens gave up laying. Worse, this was the malaria season. Of the 87 people Campbell examined that spring, 78 had malaria. Each year two, three or sometimes four children died.

Three months later, Campbell returned to check on his tower. Just after dusk, a stream of bats emerged. It took five minutes for all of them to come out. This was promising, but the tower could hold many more bats than that – at least 250,000, perhaps even half a million. Campbell knew of two roosts nearby, one in a derelict ranch house, the other in a duck hunters' shack. If he could evict the bats, maybe they would seek refuge in his tower.

Campbell had tried evicting bats before. Shouting, clapping, even hosing them down with water would dislodge them, but they always returned. He decided to try music. Campbell reasoned that because bats have sensitive hearing, tuned 'to detect the soft sonorous tones made by the vibrations of the wings of mosquitoes', they might be upset by less agreeable noises. 'A brass band suggested itself,' he wrote later. From hundreds of recordings he picked the Mexico City Police Band's rousing rendition of 'Cascade of Roses' 'on account of the large number of reed instruments and some blatant high notes of cornets'.

With the help of a friend with a phonograph, Campbell tested his bat-scarer at the abandoned ranch house. At 4 am the brassy tones of Mexico City's finest began to belt out of the building. An hour later, the bats began to return from hunting. They circled the house a dozen times, then fled for good. A repeat performance emptied the duck hunters' shack. But had the bats moved into the tower? They had. On Campbell's next visit the evening stream of bats took two hours to come out.

By 1914, local duck hunters reported a huge reduction in mosquitoes. Campbell visited the farmers and their families again. None had malaria. The farmers told him the clouds of mosquitoes had gone. They could work after dusk and their animals were healthy. Campbell attributed the mosquitoes' disappearance to his bats: 'There was nothing that could have brought about this modified condition except the great increase in the number of bats.'

Impressed, the San Antonio city council made it an offence to kill a bat, and stumped up for a municipal bat roost. In 1917, bats became protected throughout Texas. More towers sprang up – and not just as a way to defeat malaria. From the start, Campbell had promoted the idea of bat towers as a nice little earner. He had weighed a free-tailed bat's daily output and had done the sums. The Mitchell's Lake tower produced about 2 tonnes a year of high-quality fertiliser that fetched twice the price of the stuff from caves.

If Campbell's bats eradicated malaria, why isn't he one of the state's great heroes? Although malaria did disappear from San Antonio around this time, many doubt the bats had anything to do with it. 'Campbell assumed bats ate mosquitoes but his identification was suspect,' says Thomas Kunz, a bat expert at Boston University. A meticulous study of the free-tailed bat's diet carried out at Gary McCracken's bat lab at the University of Tennessee, Knoxville revealed that on average, 33 per cent of the diet is moths, 30 per cent is beetles and 15 per cent bugs. Dipterans – which include mosquitoes – make up just 2 per cent of the diet. Stomach analyses and lab tests of guano failed to turn up anything resembling mosquito remains. 'If the bats are eating mosquitoes then it must be a tiny part of the diet,' says McCracken.

Merlin Tuttle, founder of Bat Conservation International in Austin, Texas, isn't so sure. 'The bats' main food is moths but they are highly opportunistic. If there were a large number of mosquitoes then they might eat them.'

Today it's impossible to test whether Campbell's bats helped to eradicate malaria. The land has been drained, most of the mosquitoes have gone and there's no longer malaria in the USA. 'The disappearance of malaria may have been serendipitous,' says Kunz. As Tuttle admits: 'We simply don't know what happened.'

## ☀ Turn on, tune in, stand back

*It's the 1930s. You're chairman of one of Britain's gas companies. A fair slice of your domestic sales is to customers who still use your product for lighting their homes. Gas lamps may be noisy, smelly and dirty but installing mains electricity costs money, so your shareholders can be confident that gas sales will remain buoyant for a few decades yet. Except for one threat: the electrical wireless receiving apparatus invented by Mr Marconi. With the increasing popularity of the programmes being transmitted by the British Broadcasting Corporation, it dawns on you that wireless poses a growing menace to the gas industry. The public may go on tolerating the yellowish wavering light from their hissing gas mantles, but will they be prepared to pass up an opportunity of tuning in to the BBC's entertaining and instructive wireless broadcasts? The challenge, clearly, is to build a wireless powered by gas.*

What with Children's Hour, regular bulletins of news (read twice, 'first rapidly, and then slowly to enable listeners-in to take notes'), and music from Jack Payne and the BBC Dance Orchestra, in the 1920s wireless listening was all the rage. A minority of Britons were already wired up for mains electricity, but many more relied on heavy lead-acid accumulators to power their radios. These had to be taken to a local shop for recharging. This was inconvenient or, in the case of spillage, hazardous. The trams in Southampton, for example, carried notices requesting passengers not to put their accumulators on the seats.

The gas industry was well aware of what the newer energy source was doing to it. The author of an article in the *Gas Journal* of 5 July 1939 adopted a pained, almost resentful tone. 'There are, we suppose, few gas undertakings who have not suffered in greater or lesser degree the loss of valuable loads as a direct result of the all-mains electrical wireless set. Nor have our electrical friends been slow to take advantage of this "thin end of the wedge" in their sales campaigns. Indeed, it is safe to assume that the wireless set has more often than not turned the scales in favour of installing electricity where previously gas provided the bulk of the domestic services.'

Goaded into action by the unsporting determination of the 'electrical friends' to exploit their competitive advantage, the gas fightback began with a small Southampton company called Attaix. In the late 1920s it was selling a device called the Thermattaix based on the thermoelectric effect: the generation of electricity in a circuit made of wires of different metals in which the junctions between those metals are maintained at different temperatures. The greater the temperature difference, the higher the voltage generated.

The Thermattaix built up a usable voltage by connecting a large number of compound metal strips in series. The strips were housed in a drum some 25 centimetres in diameter and of similar height. Waste heat from the gas jets that warmed the hot junctions escaped through holes in the lid.

As the company literature carefully explained: 'The output of THERMATTAIX is comparatively small in Amperes, but more than sufficient to operate any modern wireless set employing modern valves of low consumption.' The gas consumed in heating the junctions was hardly going to earn the companies a fortune. But that didn't matter. What counted was the chance to use electrical devices in the home without the acid-slopping inconvenience of accumulators and, most importantly, without giving those carpetbaggers from the electricity companies a foot in the door.

The next and obvious step – to put a thermoelectric generator inside the same cabinet as the radio – was a few years coming. But by 1939 the fully integrated, gas-powered radio had arrived. The editor of the *Gas Journal* allowed himself a smidgen of triumph. 'The term "all-mains" wireless set... no longer has quite the same significance; for in future it can refer equally well to gas mains as to electrical mains.'

And this was thanks to the efforts of Henry Milnes of Milnes Electrical Engineering Company. His radios were about a metre tall, with the tuning equipment and speaker occupying the top part of the cabinet and the thermoelectric generator the lower section. This compartment was lined with asbestos sheeting as a precaution against fire. To turn on the set you lit the gas burners using a flash ignition device activated by the knob that also controlled the sound volume.

The sets sold for around £15 (this was 1939, remember, when £5 was a good weekly wage), with running costs estimated at around half that of radios using rechargeable accumulators. Gas was rising to the challenge. No wonder the *Gas Journal* felt sufficiently confident to take a gentle swipe at the opposition, enthusing that the cost of the new radio would mean that it was within the reach of those for whom it was intended: 'those who are occupying all-gas houses and who are especially susceptible to the "all-mains" sales talk of our electrical competitors'.

As if to clinch the argument, the article drew attention to a modest bonus available to users of the gas-powered radio. 'The amount of heat given off by the set is roughly equivalent to that of two small incandescent lights. This is a useful factor in assisting to warm the room.'

Radios may have been in the forefront of the gas industry's doomed struggle against electricity, but they weren't the
only domestic appliances trotted out in the forlorn attempt to
curb progress. There were gas versions of the washing
machine, the dishwasher and the gramophone. The 1920s
saw a gas-powered vacuum cleaner. It seems to have relied
on burning gas in its main chamber, then cooling whatever
products of combustion remained with a spray of water from
a tank built into the apparatus. This produced the partial
vacuum which gave the instrument its sucking power.

For obvious reasons, gas companies were happy to
display the new equipment. But even enthusiasts could
hardly have described the gas radio as a must-have household
appliance. Truth to tell, it was a big flop. One retired salesman
told *Historic Gas Times* (the newsletter for all gas enthusiasts)
that while he had gas radios in the showroom, he only
remembered selling one.

Milnes emigrated to New Zealand in the 1950s, miffed, so
he claimed, by the government bureaucracy that was
interfering with his business. But the truth was that trying to
power radios by gas had as much chance of succeeding then
as we would have today if we tried reintroducing steam
engines to drive aeroplanes.

# The human centrifuge

*'His countenance attached to saturnine blackness, the eyes, suffused
with bile, were immovably fixed on the ground, the limbs seemed
deprived of their locomotive powers, the action of the lungs, and the
circulation retarded, the tongue parched and silent, and the whole
man resembled an automaton.' In Dr Cox's opinion, his patient was
in the grip of a 'melancholy stupor'. As the director of one of the
largest private asylums in Georgian England, Joseph Mason Cox
was mad-doctor to the better off and had just the thing to rouse this*

*poor soul from the depths of his depression. Cox, one of the first qualified doctors in England to specialise in mental disorders, had invented a new sort of treatment: a human centrifuge. A spin in Cox's 'circulating swing' was said to shock the madness from a man.*

Good wine, a relaxing massage and soothing music: for Asclepiades, a Greek doctor practising in 1st century Rome, these were the best remedies for insanity. Kind and gentle treatment was far better than chains and beatings. And the best therapy of all was sleep – preferably natural, wholesome slumber rather than that induced by poppy juice or other mind-altering preparations. To encourage a better sort of sleep, Asclepiades invented one of the most enlightened pieces of medical technology: a swinging bed.

If gentle swinging was effective, then how much more might be achieved by rapid rotation? At the start of the 19th century, a radical variation on the swinging bed began to appear in asylums across Europe. However, patients treated in Joseph Cox's circulating chair found the experience anything but relaxing. Tied down and spun round at speed, they turned pale, threw up and passed out.

It was a far cry from Asclepiades's soothing swing, but it got results. Even the most disturbed patients became calm and easy to control. Cox believed any fear or discomfort was all to the good, helping to distract a patient's mind from mad thoughts. Best of all, it encouraged deep and therapeutic sleep.

Down the centuries, ideas about how to deal with the insane veered from one extreme to another: some advocated kindness, others believed that physical restraint and intimidation were more effective. Most asylums had been little more than places to lock up the mad, but by the late 18th century attitudes were changing. Cox was one of a new breed of mad-doctor. He was not a jailer or a manager of maniacs, but a medical professional who had studied mental disorders and was prepared to devote his life to investigating better ways to treat them.

The concept of swinging as therapy had gone in and out of fashion ever since Asclepiades. Towards the end of the 18th century, James Carmichael Smith, a Commissioner for Madhouses and physician to Britain's most illustrious madman – King George III – revived the notion. He suggested swinging could be used to subdue 'the nervous influence' and 'the principle of irritability' in many sorts of madness.

The idea of the human centrifuge sprang from the fertile mind of Erasmus Darwin, physician, poet and inventor. Darwin was a great believer in the healing power of sleep. But how best to induce it? Darwin's friend James Brindley provided inspiration. Although famous as a canal engineer, Brindley started out as a millwright: he'd heard that if a man lay on a millstone as it turned, he soon fell asleep. 'The centrifugal motion of the head and feet must accumulate the blood in both those extremities of the body, and thus compress the brain,' Darwin reasoned.

The same effect, he suggested, could be achieved more comfortably in a bed suspended 'so as to whirl the patient round with his head most distant from the centre of rotation'. Darwin enlisted another friend, steam pioneer James Watt, to draw up designs for a 'rotative couch', a bed attached to an arm that revolved around a vertical shaft fixed to the floor and ceiling. Darwin never built his revolving bed. It was more suited to a hospital than the parlour of a man in private practice, he said. When Cox took over his family's asylum in 1788, he was ideally placed to test it.

Cox was soon singing the praises of rapid rotation. By the time he published his *Practical Observations on Insanity* in 1804 he had considerable experience of it. Whirling his patients round at speed worked wonders, he wrote. Like Darwin, Cox believed in the restorative powers of sleep. He also believed that if you provoked some sort of physical crisis in the body, it would shock the mind back to normality, at least temporarily. Spinning certainly had a drastic effect on the body. At first the motion made patients feel nauseous; increase the speed and they vomited, then lost control of bladder and bowels. Some bled from the nose and ears; some had convulsions. Many passed out.

It was a sizeable shock to the system and invariably had a calming effect. According to Cox, even the most demented, violent patients would be left quiet and easy to control without resort to drugs. 'The slumbers thus procured differ as much from those induced by opiates as the rest of the hardy sons of labour from that of the pampered, intemperate debauchee.'

The simplest version of Cox's device consisted of a Windsor chair suspended from a hook in the ceiling and rotated with the help of ropes around the chair legs. A more sophisticated version was a bed or chair attached to an arm that revolved around a vertical shaft, much like Darwin's concept. 'The necessary motion may be given by the hand of an attendant pushing or pulling the extremity of the projecting arm, with greater or lesser force, each time it circulates, but by a little simple additional machinery any degree of velocity might be given, and the motion communicated with the utmost facility.'

By 1813 Cox was promoting spinning as a safe and effective treatment for most kinds of madness. 'No remedy is capable of effecting so much with so little hazard. In almost every case it will produce perfect quiescence, allay all irritation, silence the most vociferous and loquacious.' It was, he confessed, harder to make a madman giddy than a sane one, 'but there are very few of them who can resist the action of a continued whirling with increased velocity, especially if suddenly stopped. The shock this gives to the system and the alarm it excites is not easily conceived by those who have never witnessed it.'

Cox's chair became hugely popular in asylums in both the UK and elsewhere. In Ireland, William Hallaran, who ran the Cork Lunatic Asylum, was a great enthusiast, so much so he built a version that could take four patients at a time and spun at 100 revolutions a minute. The effect was much as Cox described: patients felt sick, threw up and later fell into a deep sleep, from which, Hallaran maintained, they awoke with their mad ideas 'totally altered'. The device, he wrote, rendered his asylum 'remarkable for its tranquillity... regularity and order'.

After a few decades Cox's chairs began to fall out of favour. Some doctors suspected they did little more than exhaust patients into submission. The treatment was dangerous – some patients died. By the end of the century the chairs had been consigned to museums. In the meantime the human centrifuge emerged in a new guise. When Austrian physiologist Robert Bárány carried out his ground-breaking research into the role of the inner ear in our sense of balance, he used a piece of equipment that differed from Cox's spinning chair in just one respect: it was called the Bárány chair. In 1914, his research won him a Nobel prize.

# 7 Reality check

Once upon a time, lambs grew on stalks, toads lived inside stones and Greek men were really gods. It is astonishing what people will believe. But even with scientific insight and the wisdom of hindsight, separating fact from fiction can be tough ...

## ☀ Toad in the hole

*Elemental creatures of darkness, toads have always been linked to the supernatural. Stone-like themselves, they were thought capable of surviving within solid rock. Even today, every encyclopedia of the 'unexplained' offers sober accounts of entombed toads. Yet despite all the stories, there's just one tangible example. It's a mummified toad nestled within a hollow flint – and the most famous specimen in a provincial English museum. Its sterling credentials are tarnished by just one thing: the man who presented it to science was Charles Dawson, the discoverer of Piltdown man.*

By Victorian times, entombed toads had become a simmering controversy that refused to go away, rather like psychic phenomena or UFOs today. Predictably, the media leapt on each new example: the living toad discovered 2 metres down in bedrock beneath a cellar in the Lincolnshire town of Stamford or the one found in a block of limestone during the excavation of a new waterworks for Hartlepool, in England's far north.

Even *Scientific American* ran a story about a silver miner called Moses Gaines who came across a tiny but plump toad lodged in a boulder.

It was all too much for the scientific establishment. When a live frog, allegedly released from a lump of coal mined in Monmouthshire, was shown at the 1862 International Exhibition in London, the letters pages of *The Times* were filled with furious demands for its removal.

So on 18 April 1901, when the Linnean Society met in London, its august fellows must have been intrigued to see the exhibit brought by one Charles Dawson. A solicitor from Sussex and a fellow of the Geological Society, Dawson was a charming man with wide-ranging interests in geology, natural history and antiquities. In his bag he carried a curious find – 'a hollow flint nodule which had been picked up on the downs at Lewes, and which on fracture was found to contain the desiccated body of a Toad'.

A small hole visible at one end must have provided the way in for a very young toad, the fellows agreed. Helpfully, Dawson had unblocked the hole, removing the chalk that must have subsequently filled it. Once inside the rock, he suggested, the toad lazily remained, content to dine on such insects that found their way in, until, too big to escape, it died entombed.

So it was curiosity, laced with sloth, that ensured the toad's fate – a salutary lesson from nature. Even better, Dawson's specimen offered a rational explanation for all those troublesome tales of toads living for millennia buried in stone. Now, at last, the truth was clear: the flint was ancient – the hollowed remains of a Cretaceous sponge – but the toad was a recent arrival. Thanks to Dawson, the mystery was solved.

A little over a week later, the toad made its second public appearance before the Brighton and Sussex Natural History Society. If anyone doubted the stories about toads being found alive in rocks, here was the explanation, Dawson proclaimed. 'Toads when small will often creep into holes in rocks and hollows in trees, and in these situations they may find sufficient food; being slothful in their habits, and capable of existing upon but little food or of abstaining from it for a long time, they are apt to remain in their snug quarters and content themselves with what insects &tc may come to them.' Besides, he reckoned, flies might have been quite plentiful, perhaps lured inside by the 'fetid and acrid exudations' from the toad's skin, or the sound of the soft scratching of its claws.

To safeguard the specimen's future, Dawson gave it to his friend Henry Willett, a wealthy Brighton collector, who included it in his generous gifts to the town's new Booth Museum of Natural History. In a small way, Dawson had made his mark in the annals of science.

And so the matter might have rested, had not Dawson gone on to 'find' the Piltdown skull. In 1953, this fossil was exposed as a deliberate fraud – with bits of a medieval human cranium married to fragments of an orang-utan's jaw. By that time, however, Dawson was long dead, and the fraudster's trail had gone cold.

Could it be that the toad-in-the-hole was a dry run for the Piltdown fraud? Not everyone agrees that Dawson was the culprit. In a frenzy of speculation, nearly every scholar in the vicinity has been accused at some time. The smart money, though, is still on Dawson. Now it looks as though he used the toad-in-the-hole to perfect a style of disclosure that would convince, for a time, the world's leading palaeontologists.

Consider the curious parallels between the two finds. In both instances, Dawson said the objects had been discovered several years earlier. The flint tomb, he told the Brighton naturalists, had turned up 'about two summers ago'. He gave the skull an even longer trajectory, dating finds to 1908 and 1911; then he waited until February 1912 to write to Arthur Smith Woodward, keeper of geology at the British Museum (Natural History) – now London's Natural History Museum.

Both times, too, the reputed finders were local workmen,

whom Dawson named, confident that in the class-ridden society of his day no one would ever bother to interview them.

In both cases, Dawson provides exquisite detail. Workmen digging gravel for road repairs found the pieces of skull at Barkham Manor near Piltdown, while the toad-in-the-hole had come from a quarry at the foot of the downs, north-east of Lewes. Its peculiar lemon shape and its comparative lightness had attracted the attention of the men, Dawson claimed. Curious, one labourer, Mr Thomas Nye, broke it open, and found the 'mummied' toad.

In retrospect, the whole thing seems suspicious. The downs are littered with hollow flints, formed around Cretaceous sponges, so Dawson's specimen would hardly be a novelty for the road menders. Equally implausibly, in the Piltdown story he tells us the workmen thought they'd found a coconut, and so decided to smash it. This yarn explained the fragmentary remains, which made disguising the true nature of the fake far easier.

With both hoaxes, Dawson's next move was to associate the specimen with someone in authority who was already a friend. For the flint, he brought in the respected businessman Henry Willett, known for his collection of Cretaceous fossils from the chalk quarries of Lewes. For his later scam, Dawson aimed higher: to validate the skull forgery, he recruited a distinguished colleague he knew from the Geological Society, Woodward of the British Museum. In 1913, Dawson was first author on their joint paper, 'On the Discovery of a Palaeolithic Human Skull', published in the society's journal. Now, at last, Dawson must have imagined, a fellowship of the Royal Society could not be far away.

Alas for Dawson, that ultimate accolade never came. He died suddenly of septicaemia in 1916, aged 52. But he lived long enough to see his skull the talk of the town, and his toad one of the most popular exhibits in Brighton.

The toad's caretaker today, geology curator John Cooper

of the Booth Museum of Natural History, is amused by the toad's enduring popularity. At first, he took it at face value, but was troubled by the fact that the amphibian looked considerably bigger in the original 1901 photograph than it does now. For it to have continued to shrink over the ensuing century, it must have only just begun to dry out in 1901. It all fits, Cooper argues. If you had this great idea – you'd seen just the right flint, and were going to concoct a hoax – you wouldn't spend 10 years drying a toad. You'd get on with it.

And just how plausible is Dawson's story anyway? Cooper began to wonder. Toads don't hang about on the dry chalk, and if the cobble had at some stage been transported to a wet stream bed, why didn't the entombed toad just rot once it died? Every fossil owes its existence to a series of improbable events, but even so, this toad was pushing it.

So how did Dawson mummify the toad? Did he inject it with alcohol, or dry it in an oven? Perhaps he pickled it. It would be interesting to test it for salt, suggests Cooper, and fitting too, for it was chemical analysis that ultimately revealed the Piltdown skull's true history. Whatever happens, the toad-in-the-hole postcards in the Booth museum shop are likely to remain bestsellers.

## ☀ Pharaoh's ears

'Three years ago, a mummy was unrolled in London, and in its hand was a small bag of Wheat. Some grains of it were sown and vegetated. Its produce has again been sown... and has produced an average of 38 ears or spikes for each grain sown. To be sold in packets of 10 grains each at £1 per packet... ' In 1843, when The Gardeners' Chronicle ran this ad, the public was crazy about ancient Egypt. And nothing was more fascinating than the notion that 'mummy wheat', grain discovered in the tombs of kings, would spring to life after thousands of years. From the start, botanists *dismissed the claims as romantic nonsense. Yet the belief in the astonishing powers of ancient seeds lingers on. In an attempt to debunk it for good, scientists at the Royal Botanic Gardens, Kew turned to sophisticated mathematical models to calculate exactly how long grain could survive in an Egyptian tomb.*

Blame it on Napoleon. When he invaded Egypt in 1798, he took along 175 scholars. Although Napoleon's army failed to conquer Egypt, his troop of intellectuals were triumphant. They 'discovered' ancient Egypt and so triggered a craze that swept the whole of Europe. Fashionable society was soon in the grip of mummy fever. By the 1840s, the English papers carried regular reports of the amazing regenerative powers of 'mummy wheat', grain discovered in tombs up to 6,000 years old.

Some of these ancient seeds produced bumper yields from fat heads with as many as seven ears. In 1846, *The Agricultural Gazette* reported the excitement among members of the Newcastle Farmers' Club when they were shown an ear of wheat grown from a seed found in an Egyptian mummy case. 'It is much more bulky than an English ear, being, in fact, seven English ears rolled into one!' This ear was more than a miracle of resurrection, pronounced the *Gazette*, it was evidence that the biblical story of Pharaoh's dream was true. 'And Pharaoh slept and dreamed the second time, and behold seven ears of corn came up upon one stalk… '

Each claim of success in raising mummy wheat brought a swift counterblast from botanists. Close examination of the seeds showed that although the outside looked intact, the embryo within was too badly damaged to germinate. Every attempt to grow authentic mummy seed failed. The sceptics argued that if anyone succeeded, then the grain they planted wasn't as old as they had been led to believe.

In one famous instance in 1840, Martin Tupper, a popular writer of the day, claimed to have raised plants from mummy wheat. His seed came from an impeccable source – none other than Sir John Gardner Wilkinson, the most eminent of English Egyptologists. Later, it turned out that Tupper's own gardener had scattered modern seed among the ancient to ward off disappointment.

There were other reasons why mummy wheat might be more modern than any mummy. For centuries, ancient tombs had been used to store grain. Mummies were often shipped to Europe packed in straw, with seed heads still attached. And Egyptian guides quickly realised that gullible tourists would pay well for grain, and kept handy supplies hidden in the tombs.

But it was the plants themselves that provided the best case against mummy wheat. The ancient Egyptians grew two main cereal crops, emmer wheat and barley. Obviously something was wrong if the seeds grew into oats or maize – species not present in ancient Egypt. When the plants did turn out to be wheat, it was usually a modern variety of bread wheat. And when the wheat was the seven-eared sort that Pharaoh dreamed of, it was invariably rivet wheat, unknown in Egypt at the time.

Despite all the evidence, the story still wouldn't die, much to the annoyance of Wallis Budge, keeper of Egyptian antiquities at the British Museum. For years, he received two or three letters a week asking if it was possible for mummy seeds to grow. No, he replied, it was not. In 1897, in an effort to dispel all doubt, he bought a model granary that had been excavated from a tomb at Thebes. The grain inside was 3,000 years old. Budge gave some of the wheat to the director of Kew, who planted it under the watchful eyes of almost the entire staff. 'They waited day after day, week after week, but no shoot of any kind appeared,' Budge wrote in a letter to *The Times*. 'At length after three months, they turned over the little plots and found that all the grains had turned to dust.'

In 1933, the press reported a new case of wheat that had sprouted from ancient seeds: this time the grain came from a 4,000-year-old tomb in India. Would Budge now admit there might be something to this mummy wheat? 'No competent botanist believes that ancient Egyptian wheat will germinate,' he declared in *The Times*. But lest anyone doubt it, he still had some seed left and would donate some to any reputable grower willing to test it.

The National Institute of Agricultural Botany took up the challenge, and reported its findings back to *The Times*. 'After the fourth day, the grains had become slimy. At the end of 16 days in test, not only had every grain completely decayed, but a thick growth of mould had spread from them to the surrounding sand.'

So why do many people, including archaeologists, still believe that these ancient seeds will grow? This is a question Mark Nesbitt, an archaeobotanist at Kew, has spent a long time pondering. He suspects that although earlier botanists debunked the idea fairly convincingly, they never provided a proper explanation for why a seed couldn't survive thousands of years if it was kept in the arid interior of a tomb.

Over the past few decades, scientists working in seed banks have discovered much about the ageing process inside a seed, and the conditions that will prolong its life. Most cereals can be stored for centuries if they are partially dried and kept at sub-zero temperatures and low humidity, although exactly how long depends on the species. With the help of artificial ageing experiments, seed-bank scientists have produced sophisticated models that predict the shelf life of a seed under given conditions.

At Kew's Millennium Seed Bank, John Dickie has modelled the lifespan of mummy wheat. Cereals all behave much the same in storage, so Dickie assumed mummy wheat would deteriorate at a similar rate to modern varieties. Feeding this data into the model, he then added information about conditions in an ancient tomb. One of the best-studied rock tombs is that of Nefertari, the favourite wife of Ramesses II, who lived in the second millennium BC. The relative humidity in the tomb is a low 16 per cent, ideal for seed storage. But even deep inside the rock the temperature fluctuates widely, ranging from a low 16 °C to 28.5 °C – bad news for seed survival.

Dickie found that if he started with top-quality seed and the temperature remained constant at 16 °C, one grain in a thousand might still germinate after 236 years. With the temperature sometimes hitting the high 20s, the grain would all be dead in 89 years. And if the seed was less than perfect to begin with …

## ☀ Last-chance balloon

*Why would a small red-and-white balloon rescued from a Victorian shrubbery in the west of England have excited weeks of feverish activity at the Admiralty in London? Because it appeared to have attached to it a message from the doomed HMS Erebus, captained by Sir John Franklin, by then missing for six years in the Arctic. If the message was genuine, it meant that Britain's greatest mariner of the day and his 130 crew – or at least some remnant of the expedition – were still alive, locked in the ice of the North-West Passage. It meant new efforts should be made to rescue them. The Admiralty hushed up the affair of the Gloucester balloon and it remains curiously ignored to this day. The message was almost certainly a hoax. But who perpetrated it? And for what purpose? And where did the hoaxer get the Admiralty-issue balloon?*

On the morning of 5 October 1851, a strange balloon was spotted floating over the cathedral city of Gloucester in the west of England. By lunchtime it had reached the outskirts of the city, where it descended gently into the shrubbery of a certain Mrs Russell of Wotton Lodge. Curious, she dispatched one of her servants to retrieve it.

The silk balloon, partially filled with gas, was less than a metre across and carried a small, soiled card declaring:

'Erebus. 112°W. Long: 71°N. Lat. September 3rd 1851. Blocked in.' That was it. Nothing more.

Everyone knew the name Erebus. Sir John Franklin was one of the great public heroes of the age. His two ships, HMS *Erebus* and HMS *Terror*, had left Britain to find the elusive North-West Passage. Franklin's expedition had sailed six years before – with three years' supplies. By now, hopes that anyone from the expedition might still be alive were fading fast.

The previous winter half a dozen rescue ships from Britain and the USA had combed the countless channels and bays of the Canadian Arctic, but found no trace of the vessels or their crews. Many would-be rescuers were convinced that Franklin and his men must be dead – frozen, starved, even, some said, eaten by Eskimos. But the public was still obsessed with the story, which filled the newspapers and became the subject of popular ballads. In the final refrain of the haunting *Lady Franklin's Lament*, the explorer's wife promises a fortune for her husband's safe return: 'Ten thousand pounds would I freely give, to say on Earth that my Franklin do live.'

And Jane Franklin, a determined and resourceful woman, would certainly have paid that. She devoted several years to berating the Admiralty into sending rescue missions and flattering a motley collection of whaling captains and adventurers into offering their services. But by the autumn of 1851, the Admiralty was increasingly unwilling to fund and organise the search.

And then the balloon turned up. The story of what happened next has been pieced together by W. Gillies Ross, an expert on Arctic exploration at Bishop's University in Lennoxville, Quebec. Delving among old Admiralty papers, Ross discovered that, while extremely reticent in public, in the weeks after Mrs Russell sent them the balloon the Admiralty embarked on a feverish investigation into the provenance of the apparent message from HMS *Erebus*.

Such a balloon, the Admiralty concluded, could have

made the journey, and the location in the message could have been correct. But as it now turns out, the ship couldn't have been iced in at the position given in the note, because it's in the middle of Victoria Island. But no one knew that then. What's more, there was no record of Franklin taking any balloons with him. Certainly he took none like the one Mrs Russell recovered.

The balloon seemed to come from a batch made for civil engineer George Shepherd, who supplied them to several of the rescue vessels that headed north in 1850. The aim was to release them across the Arctic with notes telling where provisions and assistance could be found, in the hope that Franklin would find one.

Could one of the rescuers have sent a hoax message from the Arctic? It seemed possible, though the condition of the card and the attaching twine suggested it had not made a long journey. Some suspicion fell on the officers of an Admiralty rescue ship skippered by Captain Horatio Austin which had been carrying balloons and returned from the Arctic in the weeks before the balloon's discovery. The Admiralty claimed to find no evidence that balloons were missing from the ship, though, as Ross points out, if they did, one could hardly blame them if they decided to hush it up.

Certainly, the ship's officers would have had the opportunity. But where, as any detective would ask at this point, was the motive? Ross can see none. Besides, the wording of the message does not suggest a nautical hand. The name of the ship was not preceded by the letters HMS; the words 'blocked in' were not part of naval jargon; and the longitude was given before the latitude – all contrary to naval custom. And the coordinates did not include the more precise minutes and seconds – again suggesting an amateur hand.

So who else had the opportunity? Balloons like this were not widely available. Most had been made specifically for Shepherd. He had released some during trials in London before they were dispatched to the Arctic. One of these might have been retrieved and reused. But, as Ross discovered, Shepherd also gave out several balloons as souvenirs to VIPs in attendance. He gave one to the President of the Civil Engineers' Institute, another to someone from the Royal Society – and two to Lady Franklin.

Now, Jane Franklin, alone of all the people who might have got hold of one of the balloons, had a clear motive for such a hoax. A forceful woman of 60, she had fought for the rights of female convicts while her husband was governor of Tasmania. She travelled widely on her own account, and after her husband's disappearance had orchestrated the rescue effort, almost single-handed.

Could she have committed the deed in the hope of forcing the Admiralty to renew its flagging rescue efforts? 'Such a stunt might almost be in character; she was incredibly determined, devious in some ways and could manipulate politicians,' says Ross.

The identity and motive of the hoaxer will probably never be known. But Lady Franklin must be a prime suspect. She alone had the opportunity and the motive. And she certainly had the wit and guile.

## ☀ Lamb's tales

*Not long ago, a survey of nine- and ten-year-old city kids found that 70 per cent of them thought that cotton comes from sheep. It's an easy enough mistake if you think cotton is the stuff that comes in fluffy white balls of 'cotton-wool' – especially if you've never actually seen a sheep. There's less excuse for the generations of explorers, scholars and philosophers who believed that the soft fibres from which eastern people wove fine white cloth came from the borametz, or lamb plant, a fabulous creature that was half-plant, half-animal. Towards the end of the 17th century, the worthies of the Royal Society denounced the 'lamb plant' as a figment of ancient imaginations. But where had people got the idea? The Society's top minds puzzled for a while then plumped for an object recently sent from China. It was made from the root of a tree fern, but had four legs and was covered with a sort of downy fur. This, the Society declared, was the origin of the mythical borametz. In its eagerness to nail the myth, however, the Royal Society got it completely wrong. It took two more centuries before the debunkers were themselves debunked and the origin of the borametz was finally explained.*

Rumours began to circulate in the Middle Ages. Far away in the land of the Tartars lived a thing that was neither plant nor animal, but both. It was called a borametz. None of those who told the tale had actually seen it, but they'd all met a man who had.

In some versions of the story the 'vegetable lambs' were the fruits of a tree that grew from a round seed like that of a melon. When the fruits ripened, they burst open to reveal tiny lambs with soft white fleeces that the natives used to make fine cloth. In other tellings, the seed gave rise to a lamb that grew on a stalk rooted in the ground, and lived by grazing on any plants it could reach.

The man responsible for spreading the story in Britain was John Mandeville, a 'Knyght of Ingelond' who left home in 1322 and for the next 34 years 'travelide aboute in the worlde of many diverse countreis'. His account of what he saw was the medieval equivalent of a bestseller, and was translated into every European language. In Tartary, he wrote, 'there groweth a maner of Fruyt, as though it weren Gowrdes: and whan thei ben rype men kutten hem ato, an men fynden with inne a lytylle Best, in Flesche, in Bon and Blode, as though it were a lytylle Lomb with outen Wolle.'

There's doubt about whether Mandeville ever reached Tartary, but he'd certainly read the Medieval guide books before putting quill to paper. In 1330, Friar Odoricus, who came from a monastery near Padua, wrote that 'there grow gourds, which when they are ripe, open and within them is found a little beast like unto a young lamb'. He had heard this from 'persons worthy of credit', which was good enough for Mandeville.

With each telling, the story gained new details and greater credibility. But in the 16th and 17th centuries, people learned more about the world and its inhabitants. As doubts crept in, more sceptical travellers set out in search of the mysterious lamb of Tartary. Still it eluded them, yet most came home convinced that it existed. One of these was Baron Sigismund von Herberstein, who represented the Holy Roman Empire at the Russian court in the early 1500s.

The baron had dismissed the sheep-on-a-stalk as fable until he heard it described by a 'person in high authority' whose father had once been an envoy to the King of Tartary. The envoy had seen, 'a plant resembling a lamb... It had a head, eyes, ears and all other parts of the body, as a newly-born lamb. It was rooted by the navel in the middle of the belly, and devoured the surrounding herbage and grass, and lived as long as that lasted.' The lamb also had exceedingly soft wool, and perhaps not surprisingly given its limited mobility, it was a favourite food of wolves. The story convinced the baron.

And so it went on. When anyone voiced doubts, someone else popped up with new 'evidence' of the lamb's existence. In 1605, Frenchman Claude Duret devoted a whole chapter of a book on plants to the borametz. But then, 80 years later, the great traveller Engelbrecht Kaempfer went east looking for it. He found nothing but ordinary sheep. The number of believers was dwindling, and in London the Royal Society decided it was time to kill off the borametz for good. The best way, it felt, was by showing people how the idea had begun.

In 1698, the society received a curio from China, a sort of toy animal made from a plant with a few extra bits stuck on to give it a proper number of legs. This, it reckoned, was what had started the ancient rumours. It was both plant and animal – a 'specimen' of a borametz, in fact. Hans Sloane, later founder of the British Museum, described the 'specimen' in the society's *Transactions*: it was made from the root of a tree fern, had four legs and a head and 'seemed to be shaped by art to imitate a lamb'. The four-footed fake also had a 'down of dark golden yellowish snuff colour' – a sort of golden fleece. Despite this discrepancy in the colour of its 'fleece', and the fact that tree ferns didn't grow in Tartary, the Royal Society now considered the case closed.

And so it was, more or less, until 1887, when Henry Lee, a one-time naturalist at the Brighton Aquarium, pointed out that the society's Chinese 'lamb plant' was a red herring. There was, however, a plant that had almost certainly given rise to the notion of the borametz. It grew in the East and its fruits contained soft white fleece like a lamb's. The plant was cotton.

Lee had been delving into the writings of ancient travellers while researching his books *Sea Monsters Unmasked* and *Sea Fables Explained*. As he read he kept coming across descriptions of plants that sounded far more like the prototype borametz than any hairy fern root.

The 5th-century BC Greek historian Herodotus, for example, wrote that in India 'certain trees bear for their fruit fleeces surpassing those of sheep in beauty and excellence, and the natives clothe themselves in cloths made there from'. A century later, Nearchus, one of Alexander the Great's generals, also reported that 'there were in India trees bearing, as it were, flocks or bunches of wool, and that the natives made of this wool garments of surpassing whiteness'. And in 306 BC, that reliable botanist Theophrastus added an important detail: people cultivated the plants in rows, and when the pods ripened 'the wool is gathered from it, and woven into cloths of various qualities'.

The Royal Society, Lee concluded, had failed to spot the obvious connection and had settled for something so unlikely it had to be wrong. For Lee, it was easy to see how 'a plant bearing as its fruit fleeces which surpassed those of lambs in beauty and excellence' had become a 'plant bearing fruits within which was a little lamb … '

Distorted descriptions of the cotton plants seen in India preceded the actual plants by many years. In the meantime, traders brought samples of cotton 'wool' along trade routes that passed through the Tartar lands. To those who had never seen raw cotton, this fine 'Tartar wool' looked like something that might come from the fleece of a pure white lamb. And in a world inhabited by any number of fabulous beasts, a lamb that grew from a seed wasn't so strange …

## ☀ When men were gods

*In 427 BC, the Greek city-state of Athens crushed a revolt in Myt-ilene on the Aegean island of Lesbos. The Athenian assembly decided that all men in Mytilene should be killed in punishment and dis-patched the order by the fastest means it knew – a trireme, the classic oared warship of the ancient Mediterranean. The next day, the assembly relented and sent a second trireme to call off the mas-sacre. Mytilene was 340 kilometres away and the first ship had a day-and-a-half start – but by rowing non-stop for 24 hours, the crew of the second ship arrived in time to stop the slaughter. Modern crews who tried to match this feat in a reconstructed trireme have never come close. So were ancient Athenian oarsmen supermen?*

If triremes are the stuff of legend, then so too are the men who rowed them. The sight and sound of squadrons of these war-ships closing in during the final stages of a sea battle must have struck terror into the hearts of the bravest enemy. Armed with a great bronze ram at its prow, each ship was powered by 170 oarsmen arranged in three banks, one above the other and with one man to each oar. The oarsmen could row far and fast, and accelerated to awesome speed as they prepared to ram their target. Equally impressive, they could manoeuvre their ships with agility, able to turn or back out of danger almost as fast as they moved forwards.

The heroic exploits of ancient Athenian oarsmen are not tall tales exaggerated down the centuries but a matter of record, written down by chroniclers with first-hand knowledge of the ships and their campaigns. The Greek historian Herodotus attributed one of Athens's greatest triumphs – the defeat of the Persians at the battle of Salamis in 480 BC – to the skill and physical prowess of the oarsmen who powered its fleet of triremes. Their muscular efforts and those of their successors ensured Athenian dominance in the eastern Mediterranean for the next half-century, providing the stability and prosperity that led to the flowering of classical Greek culture and the city's reputation as the cradle of western civilisation.

Greek writers left accounts of many trireme voyages, and though they describe different routes and different wars, they are fairly consistent on the matter of speed, often suggesting that a trireme crew could row at up to 7 or 8 knots (13 to 15 kilometres per hour) for 16 hours and longer. By anyone's standards, this showed prodigious stamina. So how would these men have measured up against today's top athletes?

In 2004 exercise physiologist Harry Rossiter from the University of Leeds in the UK and historian Boris Rankov of Royal Holloway, University of London, both experienced racing oarsmen, had a rare opportunity to find out what the trireme crews were made of. That year, Greece hosted the Olympic games, and the Olympic flame was to be rowed into harbour at Piraeus, near Athens, by the trireme *Olympias*, the famous reconstruction of a 4th-century- BC Athenian warship. After a decade in the Greek navy's museum, the ship was taken out of mothballs and a crew trained for the big event. 'This was probably the last time the ship would be put into the water, and so we grabbed the chance to do some experiments,' says Rossiter.

*Olympias* was the result of years of research by John

Morrison, a historian at the University of Cambridge, and British naval architect John Coates. No wreck of a trireme has ever been found, and so to design the ship they had to draw on information from literary sources, pictures on pots and marble reliefs, and indirect archaeological evidence such as the dimensions of excavated ship sheds that once housed triremes – all reconciled with the principles of physics and naval architecture. The ship was then built by the Greek navy and launched in 1987. The Trireme Trust, founded by Coates and Morrison, conducted five seasons of trials to test the design, after which the ship was consigned to the navy's museum.

The trials left little doubt that in general the design worked. Even the most practised crew, though, came nowhere close to matching the endurance of their forebears. They managed just under 9 knots in a sprint – a reasonable ramming speed – but could keep it up for only a few seconds. Over distance they could sustain a top speed of no more than 5 knots. Yet the historian Xenophon implied that even a moderately good crew could manage 7 knots for many hours. In the race to Mytilene, as recounted by Thucydides, who had himself once commanded a fleet of triremes, the pursuing ship must have averaged this sort of speed for more than 24 hours. How did triremes achieve these cruising speeds?

With *Olympias* back at sea and the latest crew well into their training, Rossiter and Rankov travelled to Piraeus to investigate the source of the discrepancy between ancient and modern performances. By measuring the metabolic rates of a group of oarsmen, they established how much energy each rower expended in powering the ship at different speeds. 'After about 4 knots the values we came up with for the metabolic demands were very high,' says Rossiter. 'It was clear that a sustained 7 knots was outside the aerobic capacity of the modern oarsmen.'

The measurements also showed that not all the power generated by the human engine went into propelling the ship

– at 7 knots, the loss is around 30 per cent. Some wastage is inevitable: friction or slippage of the oars, and even moving the oar through the air between strokes, all cost energy.

Minor alterations to the design of the ship could reduce the losses, says Rankov. One improvement would be to increase the length of each oar stroke, something that the ancient Athenians are likely to have known. New evidence has emerged since *Olympias* was built that suggests the ship is too short: it is 37 metres long, whereas a 4th-century BC trireme was probably closer to 40 metres. At the very least, the added length would increase the space between rowers, giving them an extra 9 centimetres of reach before they hit the man in front. If in addition the oarsmen's seats were skewed so they faced slightly outwards, the man in front would no longer be an obstacle and there would be no restriction on the length of their stroke.

'Our next project aims to establish whether a redesign of *Olympias*'s oar system can free up some of the "ineffective" power,' says Rossiter. 'But even if all the lost power were captured, that still wouldn't close the gap in performance completely.' These men still had to have been exceptional athletes by modern-day standards. 'Their endurance was extraordinary,' says Rankov. 'In that respect, compared to anyone you could find today, they were super-athletes.'

What is astonishing is not that men of such athletic prowess existed in ancient Athens, but that the city could find so many of them: at one point it had 200 triremes, requiring 34,000 oarsmen. What makes their achievements still more impressive is that the men of that time were small, on average only about 168 centimetres tall. World-class rowers today tend to be 190 centimetres or taller.

The Athenians, however, knew how to get the best out of oarsmen. They hired free men, paid them well and fed them well: the orator Demosthenes implies that crews that did not get proper meals became demoralised and started to desert. Equally important, they were given long, rigorous training.

Although triremes carried sails, they were faster under oar, so crews had to be fit at all times. Thucydides was very clear on the importance of training: 'Sea power is a matter of skill, and it is not possible to get practice in the odd moment when the chance occurs, but it is a full-time occupation, leaving no moment for other things.' On long voyages commanders would order the sails to be stowed and their men to row in order to get them to peak fitness. Organised races were another way in which oarsmen were prepared for war. 'It's quite clear from the texts that they knew you had to train a crew up. You couldn't just take a fleet out and expect it to perform well,' says Rankov.

Even so, it seems the ancient oarsmen had something more. Their numbers imply that unlike today's top athletes they didn't belong to a tiny elite, and no amount of training will produce a super-athlete unless the potential is there to begin with. Did they have more athletic genes? It's possible. 'Our findings throw up the intriguing suggestion that they had a greater intrinsic capacity for aerobic exercise than their modern counterparts,' says Rossiter. 'Whatever the explanation, we are left feeling distinctly inferior.'

# 8 War and peace

War, or the threat of it, invariably prompts a surge of ingenuity and unforeseen spin-offs for later generations – but who would have thought the beneficiaries would include surf bums, music lovers and Shakespeare scholars?

 ## Surf's up

*Walter Munk was never much of a surfer, but that didn't stop him from becoming a legend in the sport. An oceanographer by training, Munk has spent nearly 70 years studying how waves form, how they travel and how they break when they hit the beach. In the Second World War, he saved countless lives by helping the Allied military determine when troops could make amphibious landings without being swamped by big surf hundreds of metres from a hostile shore. After the war, Munk's methods helped surfers find the biggest waves. Today, anyone who checks out a surf forecast on the internet is drawing on his pioneering research.*

In the summer of 1942, Walter Munk went to the beach. It wasn't a holiday; Munk worked for the Pentagon and he was there to watch American troops practise for an amphibious landing in north-west Africa. The Allies were losing the war, and the invasion would be their attempt to retake the initiative.

The troops were using boats called LCVPs (Landing Craft,

Vehicle, Personnel) – smaller versions of the drop-bow boats that would later storm beaches from Normandy to Iwo Jima. They were not the most seaworthy of vessels. 'When the waves exceeded five feet, the LCVPs would swamp,' recalls Munk, now emeritus professor at the Scripps Institution of Oceanography in La Jolla, California. 'They would call it a day, and wait for another that was a little calmer.'

Munk was concerned. 'I went back and learned about waves at the landing beaches in north-west Africa. In the winter they exceeded six feet most of the time. I thought a terrible catastrophe was about to happen.'

His commanding officer dismissed his objections. 'They', the officer said, must have figured this out. Today, the 93-year-old Munk is convinced there never was a 'they'.

Everyone knew that waves were generated by distant storms, but no one knew where they came from or why, and no one had tried to make surf forecasts. Munk decided to tackle the problem on his own, puzzling out the physics of how storms generate swells and what happens as they hit a beach after crossing thousands of kilometres of open water. A month later, he took his findings back to his superiors. Again they brushed him off.

Luckily, he was stubborn. Before the war, Munk had been studying oceanography, so he took his concerns to his mentor, Harald Sverdrup, then director of Scripps and widely regarded as the top oceanographer in the USA. Previously, the two had worked on anti-submarine warfare, looking for ways to help defend North Atlantic convoys from German U-boats. Munk, however, had been born in Austria and Sverdrup had relatives in German-occupied Norway. Partly because of rumours spread by rival scientists, both had trouble maintaining security clearances in the super-clandestine field of anti-submarine warfare.

'Ironically, the US may have inadvertently won the war by denying [them] clearances and getting them out of anti-submarine warfare and into surf forecasting,' says Peter

Neushel, a historian at the University of California, Santa Barbara, and an avid surfer.

Despite his security problems, Sverdrup had the clout to get military chiefs to recognise the need for forecasts. He and Munk developed a model but they still needed to test it. 'We were desperate to see whether the method was working,' Munk says.

Then they learned that before the war Pan American World Airways had been flying seaplanes on an Atlantic test route between Bermuda and the Portuguese archipelago of the Azores. On each trip, the pilots had recorded the height of the surf. It was a treasure-trove of data.

Sverdrup and Munk hunted down old weather maps, then plugged the weather data into their model to 'hindcast' wave heights for each Pan Am landing. To their joy, they found a close correlation with what the pilots had recorded – with one puzzling exception. 'Once in a while, the observations showed a big spike which we missed,' says Munk.

Then they noticed that these spikes always occurred on Saturday nights. 'We decided they were related more to Portuguese wine than meteorological conditions,' says Munk.

Reassured, the oceanographers were able to give the thumbs-up for a landing in north Africa on 8 November 1942. This time the army listened, and 35,000 US troops went ashore in calm surf. 'I'd say we were tremendously lucky with not-so-good weather maps and not-so-good beach charts,' says Munk.

Meanwhile, Munk and Sverdrup set up shop at Scripps and began turning out entire classes of surf forecasters. Soon, their disciples were spreading out into the Pacific or teaming up with counterparts in the UK to try to figure out how to predict surf conditions in Normandy, for the long-anticipated D-Day invasion.

The British group, based at the Admiralty's lab on the river Thames, combined Scripps graduates with a team led by

George Deacon, the UK's top oceanographer. Dubbing themselves 'The Swell Committee', they recruited coastguards to collect as much data as they could about surf in the English Channel. They appear to have met some resistance from the coastguards, who saw their main mission as repelling German spies.

Eventually, the Channel model was accurate enough to persuade Allied commander Dwight Eisenhower to postpone the D-Day invasion by a day when faced with weather conditions linked to dangerously high surf.

Sverdrup and Munk's method of surf forecasting was then used in every major Allied landing in the war. 'As we learned more, our predictions became pretty good,' Munk says.

Only once did things go badly wrong. In November 1943 at Tarawa in the Pacific, landing craft got stuck on a reef 500 metres offshore. When soldiers jumped out with their guns and packs, they found themselves in deep water. Many drowned.

According to most histories, someone scheduled the invasion for a day when the tide was too low to clear the reef. But Munk thinks no one knew the reef was there. 'The major failing was not waves, but beach intelligence. The landing craft ran into an uncharted shoal.'

After the war, Munk's interest shifted to surfing, which he could watch from his cliff-top home in Southern California. Historically, surfers drove up and down the coast, looking for big waves. Sometimes they found them. Often they heard the frustrating refrain: 'Gee, you should have been here yesterday!'

Munk began to ponder where the good waves came from. By studying the way waves spread out as they travel, he was able to calculate that some came from as far as 15,000 kilometres away – further than the entire width of the Pacific. 'There was only one possible answer, which is that they came from the Indian Ocean.'

To confirm this, he borrowed a technique from radio astronomy and set up an array of wave detectors near San Clemente Island, about 100 kilometres off the California coast. The array allowed him to pin down the direction of the incoming swells, confirming that they had indeed originated in tropical storms half a globe away, then spent weeks travelling around the southern tip of New Zealand before reaching California. 'Today, every surfer in California is familiar with the fact that the waves come from far away,' Munk says. 'At that time, nobody knew it.'

Munk's one regret is that he missed the potential fortune to be made from forecasts. In the 1950s and 1960s, surfing was a small sport, pursued by a few dedicated enthusiasts. Now, surf forecasting is big business. 'The question "Is there surf?" is one a dedicated surfer will check every day via the web or real-time animated film,' says Neushel.

But there remain mysteries in the waves. Surfers know that waves come in clusters, with groups of big ones interspersed with smaller ones – but no one knows why. Nor is it fully understood how waves lose energy over long distances, says Munk. 'It's kind of a puzzle.'

## War and peace

*The words 'Maxim' and 'peace' somehow don't seem to go together. Hiram Maxim's most famous invention, the machine gun, has killed countless thousands, from Matabeleland to Manchuria and Mafeking to Mons. Yet there's no mistaking the face or the signature, and even if there were, it's spelled out on the box: 'Sir Hiram Maxim's Pipe of Peace'. Had Maxim done with death and turned peacemaker? Not exactly. As Maxim grew older he began to suffer debilitating bouts of bronchitis. His search for a remedy failed. So, prolific inventor that he was, he created his own inhaler, a glass globe with a swan-necked tube that guided soothing vapours where they were needed, deep into the furthest reaches of the throat. There was just one drawback: Maxim's meddling in medicine threatened to destroy the reputation he had earned as the inventor of the world's most efficient killing machine.*

England in the winter was a notoriously unhealthy place – cold, damp and almost guaranteed to bring on coughs, colds and more serious respiratory ailments. But to Hiram Maxim, the unwholesome winter air seemed a small price to pay for fame, fortune and a knighthood, his reward for inventing a weapon that changed the face of warfare.

Maxim came from Maine, and had been an inventor since his youth. But in 1881, at the age of 41, he left the invigorating climate of New England and moved to London to perfect his automatic gun. The Maxim gun fired 666 rounds a minute, using the energy from each bullet's recoil to eject the spent cartridge and insert and fire the next. Adopted by the British Army in 1889, the weapon first saw service in the Matabele War of 1893. In the First World War both sides used virtually identical versions of Maxim's gun. Soon almost every major military power had adopted it.

By 1900, Maxim had begun to succumb to regular attacks of bronchitis, one of the penalties of breathing the damp, smoggy air that blanketed London each winter. His bronchitis grew steadily worse, yet nothing his doctor recommended seemed to do any good. He consulted specialists. He traipsed from one mineral spa to another. But there was little improvement. 'I submitted to a very long system of steaming and boiling and taking the waters with no effect,' he wrote.

Eventually, he ended up in the south of France and placed himself in the hands of Monsieur Vos, who ran a famous and fashionable 'Inhalatorium' in Nice. At last, here was something that helped. He inhaled warm vapours dosed with menthol and pine for an hour at a time and for weeks on end. By April he was well enough to return to London. 'However, with the cold and foggy weather of the next autumn the trouble returned as bad as ever,' he complained.

So it was back to Nice. This time, Maxim did more than inhale. He listened to doctors and fellow patients, trying to learn as much as he could about both his sickness and Vos's apparatus. When he returned to London, he bought some glass tubing and began to experiment with an inhaler of his own.

Doctors knew that warm vapours were good – and that vapours laced with pine essence were better, for pine worked as an antiseptic. But the design of most inhalers, Maxim insisted, meant that almost all the healing vapour was absorbed by the lining of the mouth and never had a chance to work on the inflamed and painful membranes of the throat.

His solution was to design an inhaler that delivered the warm vapour exactly where it was needed, by the simple device of a dent about 5 centimetres from the end of the tube. With the teeth placed in the dent, the opening of the tube was close to the back of the mouth. 'By making a mouthpiece of such a shape that the vapours were introduced directly into the throat instead of medicating the inside of the mouth I found that my simple device was much more effective than the very elaborate machinery of Mr Vos,' said Maxim.

Convinced he was onto something, Maxim had more inhalers made and gave them away to fellow sufferers. As word of their effectiveness spread, demand grew. Eventually hundreds of thousands of them were sold.

Maxim was not content with simply delivering the remedy to the right place. He also provided what he thought was a far better form of medication. Although pine essence brought relief, it was liable to cause a tickle in the throat when the patient began to inhale – a tickle that could set off a cough, which doctors feared might strain an already weakened chest. Maxim wanted the power of pine without its irritant effect.

At the time, doctors in France added a small amount of cocaine to the inhalant, which numbed the throat and allowed a patient to inhale more concentrated pine vapours. Maxim considered cocaine a poison and swapped it for essence of wintergreen, a plant from his home state of Maine. It 'really benumbs the mouth and throat just as cocaine does, only in a less degree. By mixing a small quantity of the oil of this plant with pine essence, the vapours may be inhaled very strong without producing coughing.' He called his concoction Dirigo, from the Latin meaning 'I guide' – which just happened to be the motto on the Maine coat of arms.

Maxim's Pipe of Peace was intended for long and regular use at home, with three or four sessions of 15 minutes or more a day. For rapid relief and for travelling, Maxim made a pocket-sized inhaler ready charged with crystals of menthol, another soothing and antiseptic compound.

Despite the popularity of the device, Maxim's friends were worried that it could damage his standing as a scientist. 'Some of my friends not altogether unconnected with the gun business have told me that I have ruined my reputation absolutely by making a medical inhaler, and a scientific friend has written me deploring the fact that one so eminent in science as myself should descend to "prostituting my talent on quack nostrums",' he grumbled in his autobiography. 'It is a very creditable thing to invent a killing machine, and nothing less than a disgrace to invent an apparatus to prevent human suffering.'

Maxim couldn't really have cared less. 'I suppose I shall have to stand the disgrace,' he said. 'If I had not found the means of cure,' he added, 'I could not live on this side of the Atlantic at all.'

# ☀ Their Führer's voice

*September 1944 was not the ideal time to be making music in Berlin. The Allies had landed on the French beaches, had liberated Paris and were closing in on the German capital. Allied bombers were pounding the city and Hitler had given the order to launch the first V-2 rockets at London. Although no one outside Germany knew it, Hitler was already using another secret weapon – the first hi-fi tape recorder – to make the Allies think he was a target for bombing raids in one city, when really he was safe in another. In the midst of this maelstrom, conductor Herbert von Karajan was holed up in a Berlin radio studio, using Hitler's secret tape system to make experimental hi-fi stereo recordings of a German orchestra that still sound stunning today.*

No one has ever been able to agree on who invented television, so it won't come as much of a surprise to learn that no one agrees who invented stereo sound recording, either. One thing is certain, though. The names Helmut Kruger and Ludwig Heck are seldom mentioned, although the stereo recordings they made in war-torn Berlin 67 years ago sound remarkable even by today's audio standards.

The story begins in peacetime, at the Funkausstellung or Radio Show in 1935. Held in Berlin every two years since 1924, the show has traditionally been the launch pad for new electronic ideas. In 1935, the big news was the Magnetophon, a tape recorder made by AEG Telefunken, with plastic tape from the BASF division of German chemical giant IG Farben.

The idea of magnetic recording was already old. Watch the 1934 British movie *Death at Broadcasting House* and you will see the murderer exposed by an accidental recording made by BBC engineers on giant reels of rapidly moving metal strip. But the idea of capturing sound on thin plastic tape coated with magnetic oxide was completely new.

The Magnetophon's sound quality was poor, though. The

constant direct-current 'bias' signal needed to make the magnetic particles more sensitive to weak microphone signals blanketed the recorded sound with hiss. Getting rid of the hiss, by muffling the high frequencies, made the music sound dull and lifeless. The effect was all too evident on recordings made by BASF engineers when Thomas Beecham and the London Philharmonic Orchestra gave a concert at the music hall in BASF's headquarters at Ludwigshafen in November 1936.

All that changed in 1940, thanks to a lucky accident in a Berlin lab. Walter Weber and his boss Hans Joachim von Braunmuhl were engineers at the RRG – the German equivalent of the BBC. One day, they were tweaking the circuitry of an AEG tape recorder in an attempt to improve the sound quality, and they adjusted one control too far. The circuitry went into positive feedback – like the so-called 'howl round' heard when a microphone is feeding sound signals to a loudspeaker, which then feeds back the sound to the microphone, and so on. The tape recorder circuit oscillated at several tens of kilohertz, producing whistle tones too high-pitched to hear. But the signals shook up the magnetic particles on the tape, making it super-sensitive to weak microphone signals.

The AEG machine was suddenly able to record sound with a quality very close to today's FM radio and better than most MP3 players. Von Braunmuhl and Weber filed for patents, AEG bought the rights and hi-fi tape recording was born. The head of AEG said late in 1940: 'The use of high-frequency bias has opened up for us the possibility of making the tape recorder, with one blow, into an acoustically outstanding instrument.' How right he was. All tape recorders have used high-frequency (HF) bias ever since.

The breakthrough was good news for Adolf Hitler. From the early 1940s he was able to record a speech on tape in one city, for playback and transmission in another part of Germany. The Allied engineers who continually monitored

Germany's broadcasts in the hope of bombing Hitler mid-speech were fooled because there was none of the hiss expected from tape, nor any of the cyclic noise of disc recordings. As far as they were concerned Hitler was there in the studio, speaking live to the nation.

The Allied engineers were even more puzzled when in 1943 the RRG began to broadcast live orchestral concerts in the middle of the night. One of those engineers was John Mullin, a liaison officer with the US Signal Corps who was stationed in the UK ahead of D-Day. He realised there were no musicians in the studio: what he was hearing was recorded music captured by a new recording system far better than anything previously available. But he had no idea what it was and how it would change his life.

Berlin's radio station was based at the Haus des Rundfunks, or Broadcasting House, with a large hall that was used for broadcasting live symphony concerts. RRG had started tape-recording music in the hall in 1942 for broadcast from transmitters around the country, and the following year sound engineers Ludwig Heck and Helmut Kruger experimented with one of the HF bias Magnetophons. They added a second magnetic head so that two tracks could be recorded in parallel.

Kruger then made test recordings of concerts in stereo without telling the musicians. He hung three microphones a metre above the orchestra, one to the left of the stage to catch the violins, one on the right for the basses, and one in the centre, 2 metres behind the conductor.

The modified Magnetophon ran the tape at 77 centimetres per second to get the best possible quality. The sound from the side microphones went to the two separate recording heads, and the central sound was spread evenly between the two channels – in much the same way that concert broadcasts are recorded today.

'By the end of the war, we had about 200 to 300 excellent stereo recordings which were stored here in the bunker,'

Kruger recalled in 1993 when the international Audio Engineering Society honoured him at a convention in Berlin.

In 1945, at the end of the war, most of the RRG sound archive vanished when the Russians took over the building.

But five stereo recordings survived. Two came from the Polish town of Koscian, where the German army had built a military hospital for war invalids. The RRG laboratories had moved there too. The lab director was von Braunmuhl and he organised concert evenings for the hospital doctors at which he played the latest recordings from Berlin on a stereo Magnetophon.

At the end of the war, two of these tapes – including an extraordinarily high-quality stereo recording of Anton Bruckner's Symphony No. 8 in C minor played by the Orchester der Berliner Staatsoper under Herbert von Karajan on 29 September 1944 – were shipped back to Berlin in private luggage and given to Kruger.

Three more stereo tapes were later found in Moscow. A Russian officer had given them to the state archives, along with 1,500 mono tapes. In 1991, after the Berlin Wall fell and the Soviet Union broke up, the tapes were returned to the RRG's successor, the Sender Freies Berlin or SFB.

Engineers from the SFB and the Audio Engineering Society copied the original analogue recordings to digital tape and dubbed excerpts to CD. No electronic doctoring whatsoever was used. The sound quality, especially of the September 1944 tapes, never ceases to amaze.

That is not quite the whole story, though. In 1945, Mullin was sent to Frankfurt to investigate tales of a 'death ray' station built by the German army to bring down Allied aircraft with high-power radio signals that disrupted their ignition electrics. While there, he visited the local radio station and found several HF bias Magnetophons with BASF tape. Suddenly he understood how he had heard broadcasts of recordings that sounded live.

Mullin shipped the recorders back to the USA, and with

financial backing from Bing Crosby, who wrote him a cheque for $50,000, he worked with US electronics company Ampex and chemical company 3M to recreate the German technology. Crosby started pre-recording his radio shows in 1947.

Ironically, because Mullin took his Magnetophons from Frankfurt rather than Berlin, they were mono machines. The world had to wait until 1949 for American and British engineers to catch up with Germany and tape-record in stereo.

# ☀As Shakespeare liked it

*Venture inside the Folger Shakespeare Library, built by oilman Henry Clay Folger in Washington DC to house his immense collection of rare books, and you step back in time: the walls are panelled in dark wood and stained glass windows filter light upon hushed readers and ancient tomes. But hidden in a store room between a stepladder and a pile of toner cartridges sits a contraption once described in newspapers as 'a bulky, box-like monster about six feet tall with a row of toggle switches, flashing lights and a generally sinister appearance'. It changed the way we understand old books – and it might not exist were it not for dedicated astronomers, some borrowed children's toys and a few well-aimed beer cans.*

Modern literary scholarship began with bombs – or, at least, with a problem those bombs presented. For while it's one thing for warplanes to hit a target, it's quite another to measure the damage they inflict. And so it was that Shakespeare scholar Charlton Hinman, who joined the US navy as a cryptographer in 1941, heard an intriguing idea that military intelligence was considering. What if you ran before-and-after aerial reconnaissance photos in quick succession, creating a primitive motion picture? Any change between the two would be interpreted by the brain as movement, with bombed gun emplacements rather appropriately appearing to shake violently. But an intractable problem arose: how could you take before-and-after photos from precisely the same spot in the sky? The scheme was shelved.

The idea of running two similar but subtly different pictures in quick succession to look for telltale flickers of change was not an entirely new one. In 1904 the German instrument-maker Carl Pulfrich had invented the 'blink comparator', which flickered images fast enough for astronomers to spot changes in photographic plates taken on different nights. Subtle differences, such as a passing comet or asteroid, would cause a telltale wobble in the image. The blink comparator helped astronomer Clyde Tombaugh to discover Pluto in 1930. The same concept, brought to Hinman's attention during his navy days, would eventually revolutionise a very earthbound task – literary scholarship.

As a doctoral student studying Shakespeare at the University of Virginia in the late 1930s, Hinman pored over early editions of the Bard, determined to pin down 'what Shakespeare actually wrote'. It was a daunting task, as virtually no two early copies of Shakespeare are identical. In one copy of the 1623 *First Folio of Shakespeare*, Laertes yells out at Ophelia's graveside 'O treble woe… ' But in another he says 'Oh, terrible woer'. These should be two identical books: the same line, the same printer, the same edition. What was going on?

The explanation lies in the chaotic process of publication in Shakespeare's day. Printers corrected their pages during the print run, sometimes more than once. But they were loath to discard uncorrected and half-corrected pages, and because printers tended to haphazardly jumble and stitch their finished pages together, no two books had exactly the same combination of corrected and uncorrected pages. The result presents a riot of variations that make it nearly impossible to determine what the 'real' text was at all.

'To everyone else, they will seem rather inconsequential

– "Eureka! A missing comma!"' says Ian Gadd, an expert on the writings of Jonathan Swift. But for literary scholars like Gadd and Hinman, tiny variants reveal the origins of each page. For centuries, there was only one way to untangle these texts: manual collation, or going back and forth from one copy to another to check each word and punctuation mark. It's an agonisingly slow process dubbed 'the Wimbledon method' for its endless back-and-forth movements of the head. Little wonder that Hinman's doctoral project of manually collating and analysing multiple copies of *Othello* took him years.

What then, of his goal to collate all of Shakespeare's plays? Hinman estimated it would take at least 40 years. But, emerging from the navy to teach at Johns Hopkins University in Baltimore in 1945, the failed attempt to assess bomb damage nagged at him. What if the concept of a blink comparator was modified? What if you were comparing not planets or bombing runs, but books?

An inveterate tinkerer, Hinman set about cannibalising parts for his project, and later recalled that he had procured 'a pair of ordinary microfilm projectors (scavenged from the navy), some pieces of a wooden apple box (abstracted from a trash pile), some heavy cardboard (begged from the Folger bindery), and parts of a rusty Erector set (more or less hijacked from the small son of a close personal friend)'. Add mirrors and an eyepiece, and he was ready to succeed where military intelligence had failed.

It didn't work. The problem was that he wasn't actually comparing books: he was comparing microfilms of books. Microfilm was prone to smudges and scratches and Hinman's blinking motion picture was bedevilled by phantom shakes. The only way to avoid the problem was by using the priceless old books themselves. By 1949, Hinman had completely redesigned his machine. The new model, a gigantic sheet-metal beast, resembled a cluttered lab table with a ventilation hood and now included a system of blinking lamps and mirrors arranged around books on velvet-covered wooden racks.

It wasn't a perfect arrangement. If you placed the high-intensity lamps too close, you would burn the books. But libraries around the world eagerly bought Hinman Collators from Hinman and his builder Arthur Johnson, and the results were dramatic. Hinman himself carried out a life's work in no time, collating Shakespeare's *First Folio* of 1623 in just 19 months. Other scholars, working on everything from Nathaniel Hawthorne's *The Scarlet Letter* to Mark Twain's *Tom Sawyer*, published a flood of newly definitive editions of classic literary works. By the 1970s most major universities had a Hinman Collator, as did the British Museum and the US Library of Congress. Even the CIA bought one, though it wouldn't say what for.

Mechanical collation can be curiously disorientating. Experienced users, viewing pages as a whole image rather than reading the words on them, run their eyes in an S-pattern down the page in a matter of seconds, much faster than they could read the actual text. 'I enjoyed it immensely,' says Twain scholar Sidney Berger, 'partly because I was working alone in a dark room and could shut out the rest of the world. It's a meditative experience, once you get used to the Zen of the thing.' As the lights in the machine blink hypnotically, the trained eye quickly spots a flickering and nearly subliminal phantom text that flickers just within the edge of perception. View dual copies of Act 3, Scene 1 of *Two Gentlemen of Verona* in the Folger's collator, for instance, and the phrase 'thrice in that Article' seems curiously out of phase. Looking at one of the books reveals that it has an extra word: 'thrice in that last Article'.

Working with mirrors and flashing lights in a library still brings researchers strange looks. But collators like the Folger library's are becoming a rare sight, supplanted not by computers but by mirrors. Instead of relying on the perception of motion, newer portable 'optical collators' use the same stereoscopic principle as a child's Viewmaster. The brain naturally takes two separate images, one from each eye, and combines them to create perception of depth. By lining up mirrors and positioning your head just so, you can train your left eye on one book, your right eye on another; variations in the merged image appear to float off the page.

'It's a strange sensation,' admits Carter Hailey, the inventor of the collator's latest and simplest refinement into a two-mirror system. 'The text becomes topographic in nature.' Even pages that have been reset with identical text but slight variations in spacing – a difference invisible to the naked eye – become immediately apparent. 'The whole page leaps up in a jumble,' says Hailey.

Hinman himself never saw his invention gathering dust; he died in 1977. And though he and Johnson tried selling collators to banks to detect forgeries and to pharmaceutical companies to catch misprinted labels, neither made much money from their machines. Theirs was a labour of love, subsidised by a more profitable invention of Johnson's called the Targeteer – a poor man's trap-shooter that tossed beer cans into the air for soused gun-owners to blast away at. It's all a long way from discovering planets and analysing bombing runs. But then again, it wouldn't be the first time that a few beers have helped coax out an old writer's secrets.

# Secret service

*They look like ordinary pencils. And they write like ordinary pencils. It's the green paint that's the giveaway. During the Second World War, with paint in short supply, most of the pencils leaving the Cumberland Pencil factory had a natural wood finish. Just a few were green, a difference that marked them out as a special line. Snap one open and all is revealed. Hidden inside is a map and a tiny compass. The pencils, issued to British airmen flying over enemy territory, were one of the secret gadgets thought up by Charles Fraser-Smith, the shadowy civil servant who became the model for Ian Fleming's 'Q' in the James Bond stories. Like most of Fraser-Smith's ingenious devices, the pencils were made by a well-known manufacturer: the Cumberland Pencil Company was the oldest in the country. By using established firms, the man from the ministry was able to tap the ingenuity of a whole band of engineers and designers, and he ended up with a product bearing a well-known name that wouldn't arouse suspicion if it fell into the wrong hands.*

At 5.30 pm, Fred Tee picked up the folder with his papers in, put on his trilby and headed for home. Like most of the 100-strong workforce at the Cumberland Pencil Company, he was a local man and lived just a few minutes' walk from the factory in Keswick. As soon as it grew dark, Tee, the factory's youthful technical manager, set off back to the works and quietly let himself into his laboratory through the back door. This was the fifth night in a row that he and his fellow managers had met after work to do a spot of moonlighting.

This was 1942, and Britain was at war. Tee and his colleagues had been asked to produce a special type of pencil: it must have a secret compartment just large enough to hold a tightly rolled map and a tiny compass. In the interests of security, only the managers were in on the secret, sworn to silence by the Official Secrets Act.

Tee and the Cumberland Pencil Company had been commissioned by a mysterious man from London who claimed to be a civil servant from the Ministry of Supply's Clothing and Textile Department. He was Charles Fraser-Smith, a fixer whose real job was to supply equipment and gadgets for MI6, MI9 and the Special Operations Executive – everything from miniature cameras to surgical saws, edible notepaper and forged foreign currency. He was always on the lookout for novel ways to hide equipment that would help downed airmen avoid capture, prisoners of war escape, and secret agents get their information safely back to Britain.

He was the original 'Q' immortalised in the James Bond movies.

Fraser-Smith was bombarded with requests for devices with secret compartments, and conjured up shaving brushes, pipes and pens, golf balls and even shoelaces that concealed escape equipment. His strategy was to approach a well-known firm that made a suitable object and ask if they could make a version with some unusual features. Across Britain, designers and engineers took up the challenge.

So when Fraser-Smith needed a pencil with a secret compartment he visited the oldest and best-known manufacturer in the country, the Cumberland Pencil Company. Was it possible, he asked, to make a pencil that would hold a tightly rolled map, about 12 centimetres long, plus a compass – without anyone noticing? A pencil was a standard piece of navigation equipment, making it an ideal place to hide escape gear.

As technical manager, Tee worked out how to make the pencils without the rest of the tight-knit community finding out about them. There were six separate operations in producing a pencil – first making the leads, then gluing them into grooved cedar-wood slats, shaping the pencils and embossing them with the company name, before packing them into boxes. Although it would have been easier to create the hiding place early in the process, Tee decided that the extra step should be done right at the end to ensure that none of the workforce realised what was going on.

After hours and at weekends, Tee and his fellow managers crept into the factory, took a box of finished pencils off the shelf and carefully drilled out the insides, leaving a short stretch of lead-filled pencil at the working end. The next job was to slide in the map, fix the metal ferrule to the end, slip in a tiny brass compass and glue the rubber back on top. At the end of the job, the pencil looked just as it had at the start.

The maps and compasses arrived secretly at the factory, ready to be inserted into the pencils. The compass was one of

Fraser-Smith's early successes. Almost his first job had been to supply a fountain pen with a miniature compass inside. He tackled the problem of the pen first, and once he had established that Platignum could make one with a suitable hiding place, he set about finding a compass to fit it. 'No compass as yet existed that was small enough to fit into the tiny aperture,' he wrote in his autobiography. But in London he discovered 'a couple of back-alley brothers', the Barkers of Clerkenwell, who were making large compasses for the Navy. He asked them to make something 'smaller than they had ever seen or heard of'. They did, and over the next few years the miniature compasses turned up inside pens and pencils, in battledress buttons, hairbrushes and even in place of fillings in airmen's teeth.

The next component was the map. Fraser-Smith toyed with handkerchiefs printed with invisible ink that would emerge when soaked with urine. These were too bulky to hide inside gadgets, so he had to think of something else. The maps in the Cumberland pencils were printed on a fine, non-rustling tissue paper made specially for the job, then rolled around a soft wire which was folded over at the tip to secure the paper. Three cotton ties ensured the map stayed tightly rolled and no more than 3 millimetres in diameter. There were four maps, which were fitted into a series of pencils numbered 101 to 104. Pencils labelled 101 held a general map of Germany. The other three concealed larger-scale maps of different sectors of the country.

So did any downed airmen or prisoners find their way home with the help of a Cumberland pencil? Fraser-Smith was certain his gadgets saved lives and helped people get home, but there were no official records. Officially, he didn't exist. In their remote and tranquil setting in the Lake District, Tee and his colleagues would never find out. Their pencils didn't exist either.

# 9 Make do and mend

Run out of money? Stranded in the middle of nowhere? When needs must, it's amazing what you can conjure up from the most unlikely ingredients – a computer that models the British economy, a giant radio telescope and even a fully-functioning rainforest.

## ☀ Liquid assets

*Take a pile of Perspex tubes, a few levers and pulleys and the wind-screen-wiper pumps from an old wartime bomber. Add a brilliantly ingenious mind, a bucketful of water, and what do you have? A computer that can model the flow of money around an entire nation. If the government raises taxes or the public goes on a spending spree, then the bizarre machine cobbled together in a garage by one-time crocodile hunter Bill Phillips shows precisely what will happen to the country's savings and investments. At a quick flick of a switch the strange contraption can reveal the wisdom of increasing government spending or the folly of cutting interest rates.*

Plumbing was something Bill Phillips was good at. Economics he found harder. As a student at the London School of Economics in the late 1940s, Phillips concentrated hard as his teachers tried to explain the latest theories. But money moves in mysterious ways, through a tangled web of taxes, savings and investments, imports, exports and a whole assortment of other variables – all inextricably tied together. How much easier it would be to understand the convoluted workings of the nation's economy if you could see what happened when the government finally gave pensioners a few pounds more or froze the tax on petrol. Phillips was convinced that with a nifty bit of plumbing he could show instantly the effects of a penny more here or an extra few per cent there.

Even in 1946, when British universities saw an influx of ex-servicemen of all ages and backgrounds, Phillips was unusual. The New Zealander had just emerged from three and a half years as a Japanese prisoner of war, a half-starved, chain-smoking war hero of 32. He had grown up on a dairy farm where ingenuity was taken for granted. His mother installed running water in their house. His father diverted some of it to generate electricity to light the milking shed.

Bill had similar talents and at 15 he became an apprentice electrician. But it wasn't long before the travel bug bit and he was off to Australia, where he picked bananas, mined gold and hunted crocodiles – all the time learning to be an engineer by correspondence course. In 1937 he arrived in London, travelling via Japan and the Trans-Siberian Railway.

When war broke out, Phillips joined the RAF. He was captured by the Japanese in Java. For the next few years his ingenuity and engineering skills proved invaluable. He made a secret radio and picked up news of the first atomic bomb. He also invented a sort of immersion heater to allow the prisoners to make themselves a secret cup of tea before turning in each night. 'The result was that when some 2,000 cups of tea were suddenly brewed, the lights of the camp dimmed alarmingly… the Japanese were mystified by this dimming of the lights every night at about 10 pm,' wrote fellow prisoner Laurens Van der Post.

Back in London in 1946, Phillips signed up to study sociology at the LSE. He barely scraped a pass – a miserable result his professors blamed on his nicotine addiction. Phillips was so hooked he kept dashing from the exam room for yet another cigarette. But it wasn't only that: Phillips had been sidetracked from his sociology studies by his interest in economics and had spent too much time sketching out plans for his hydraulic model of the economy.

In a note to one of the LSE's economics teachers, he wrote: 'I have great difficulty in trying to understand your lectures. I know something about plumbing and have tried to sketch a hydraulic model... Could you please comment on it?' Surprisingly, Phillips wasn't sent packing. He was sent instead to see James Meade, professor of commerce, who was keen on mechanical devices. Meade was intrigued and told Phillips to go away and prove his idea would work. All that autumn, Phillips beavered away in his landlady's garage in south London, constructing a monstrous edifice of tubes, valves and sluices.

In November 1949, Phillips unveiled his creation before a sceptical audience at the LSE. He poured 'cash' in at the top – coloured red for better visibility – and turned on the pumps. Money gurgled around the pipes, cascaded over sluices and filled tanks. As the water levels settled, the pulleys turned and a pen plotter traced the results. To everyone's surprise, the weird machine worked.

The machine might have looked like a bizarre piece of plumbing but it was an analogue computer which accurately modelled the effects of a whole range of factors on the total national income. The movement of money was represented by the water flowing round the Perspex pipes and the accumulation of money by water collecting in tanks. As water flowed through the machine, the stream split, with savings gurgling round one loop, taxes sloshing along a separate pipe and imports trickling through another. The operator could simulate a tax cut, say, or an increase in government spending, by opening and closing valves and raising and lowering sluices. It was soon obvious if the policy led to a stable economy or total chaos.

Even the satirical magazine *Punch* was impressed. Bemoaning the extent of people's ignorance about all things financial, it recommended installing what it called the 'financephalograph' in every town hall in Britain. 'The machine is taller than the man in the street and wider and heavier and much, much cleverer... using coloured water (a convenience denied the man in the street) it reacts obediently to every morsel of economic information communicated to it, and records, with its mechanical pens on its calibrated charts, the subtle impact of a slump in the secondhand ship market, the slightest hint of a boom in soap, emery wheels or white fish.'

The model did have a few teething troubles. Inflation was sometimes a problem, just as it is in the real world. If inflation rose too high, water squirted out through a hole, leaving a pink puddle on the floor. Tired of mopping up, Phillips built new, improved models – 14 in all. Some went to British universities, others were calibrated in dollars instead of pounds and ended up at Harvard, the Ford Motor Company and the Bank of Guatemala.

The financephalograph was a fantastic creation. It earned Phillips a lowly job at the LSE. His next idea – the Phillips Curve – brought him a professorship and an international reputation as a brilliant economist. The Phillips curve demonstrates how wage hikes lead to lengthening dole queues – which is what he is remembered for today. As electronic computers improved – and punch tape gave way to figures on screens – his marvellous machines were retired. Pink puddles were consigned to history and money never flowed like water again.

# ☀ Buckets of bullion

*When HMS* Thetis *was wrecked off the coast of Brazil in 1830, a hoard of gold, silver and 'other treasure of various descriptions' sank to the bottom of the sea. The news was greeted with consternation in Rio de Janeiro. The treasure belonged to the city's British merchants, who had trusted the Royal Navy to see it safely home to England. Now it was gone and the Navy's commander in Rio ruled out any prospect of retrieving it. Everyone who knew this stretch of coast agreed that salvage was impossible. Everyone, that is, bar one.*

On 4 December 1830, the 46-gun frigate HMS *Thetis* left Rio de Janeiro bound for England. In its hold was a fortune in gold and silver bars and assorted coins worth $810,000. Held up by light winds and fog, the next evening the *Thetis* still hadn't passed the Cape Frio peninsula, some 120 kilometres east of Rio. Because of the delay, the ship would have to negotiate this treacherous stretch of coast at night, so the captain set a course that would take him well clear of the cape.

At 8 pm the watch changed. 'Almost immediately the lookout man called out "Breakers under the bow" – immediately followed by the shout "Rocks above the masthead".' These were the cliffs of Cape Frio Island, a wild and uninhabited dot off the end of the peninsula. The compass needle had lied. The instrument was either faulty or had been misled by the magnetism of the local rocks. Under full sail, the ship ploughed into the cliffs at speed. The masts splintered and fell, killing and wounding some of the crew. The deck was choked with broken spars, rigging and sails. It was pitch black and raining hard.

But the ship was still afloat. The crew tried to push away from the rocks but the ship swung back and pounded into the cliffs. Now it began to leak. The ship's small boats were all broken to pieces. Some men made a desperate leap for a ledge on the cliff. Forty made it. Others were crushed between the ship and the rocks. But the ship wasn't done for yet.

The waves swept the *Thetis* along the coast and into a notch in the cliffs. There the swell smashed it repeatedly onto submerged rocks until it sank. Incredibly, only 28 of the 300 men aboard died in the disaster. The men who had first leapt to safety had scrambled round the coast to the cove, where they rigged up ropes and slings and rescued the survivors. When word got back to Rio five days later, the Navy sent ships to pick up the men. As for the treasure – Rio's merchants would have to forget it.

Thomas Dickinson, captain of HMS *Lightning*, wasn't so sure. And if he could find a way to salvage the cargo, he was convinced that fame and fortune would follow. He made some enquiries. How deep was the water? How bad were the currents? How tall were those terrible granite cliffs? He didn't doubt it would be difficult and dangerous but he reckoned 'it was practicable, or at the least that it was worth the trial'. He would need a diving bell, air pumps and lifting gear. When he couldn't find what he wanted, he designed his own. What followed was one of the most spectacular salvage operations ever.

Dickinson planned his salvage operation meticulously. His idea was to string cables across the inlet – by now called Thetis Cove – and suspend a diving bell over the wreck. Divers could hunt for the hoard from the safety of the bell.

Diving bells at this time were little more than an iron shell open at the bottom and filled with air pumped in from above. Dickinson couldn't find either a bell or an air pump in Rio. 'After much anxious consideration, it occurred to me that it was possible to make such an instrument of iron water tanks, strengthened with bars of iron,' he wrote. He sketched out some designs, had models made, and convinced the Admiralty that he had a chance of success.

Dickinson took on a Mr Moore, an English engineer who agreed to help in return for a share of the proceeds. Under

Moore's watchful eye, Navy armourers cut and joined two heavy iron water tanks taken from one of His Majesty's warships. They reinforced the bell by riveting thick iron bars over the top and the edges. Inside, they provided a seat and footrest for two divers and hooks for their tools. 'It was lighted by six patent illuminators … and these rendered it so light that a person might see to read at a depth of many fathoms.'

Dickinson also had an air pump made, but finding hoses to carry the air into the bell was another matter. In the end, he cannibalised hoses from his own ship and made them watertight by coating them with tar and 'bandaging' them tightly with tarred strips of canvas.

The *Lightning* reached Cape Frio Island on 30 January 1831. Dickinson immediately ditched his plan to rig a cable across the cove in favour of working from a derrick fixed to the base of the cliffs and angled out over the wreck. First, though, he had to find quarters for his men. The seaward side of the island offered nothing but sheer cliffs, but on the landward side there was a sandy beach where the men built themselves a village with grass and wood huts. They lived there for more than a year.

Despite the white sand and profusion of exotic flowers, the men soon discovered this was no tropical paradise. When they returned to their village at night, they fell into beds soaked by the rain that poured almost unhindered through the walls and roofs of their huts. They suffered from colds, chesty coughs and rheumatism. Then there were the biting insects – swarms of mosquitoes, fleas and 'jiggers' that burrowed into the skin and left gaping sores. Opossums stole their food. Snakes as fat as a man's thigh infested the thatch. When it wasn't raining, the wind drove sand through every crack where it 'mingled itself with both victuals and drink'. When the sun shone, the men were blinded by the glare from the snow-white sand.

The carpenters set to work on the derrick, piecing it

together from spars saved from the *Thetis* and timber stripped from the *Lightning*. By March, the derrick still wasn't ready and Dickinson was anxious to see some sign of treasure. He had a second, smaller diving bell made that could be lowered from the back of a launch and on the last day of March Dickinson caught his first glimpse of gold. 'A tally board floated up with cheering words written on it,' he wrote in his account of the operation. The men had spotted some dollars among the debris. 'When they came up with their caps full of dollars and some gold they were received with three as hearty cheers from all of us in the cove as were ever given.'

In the next few days, the men found more gold and coins. Then bad weather stopped work. There were other frustrations. The derrick was too short and had to be extended. Eventually, the carpenters fitted together 22 pieces of wood to form a 50-metre spar. By early May the crew had the derrick in place and built a platform for the pumping gear. From now on, when the weather was fine, both bells were kept working, the big one suspended from the derrick, the small one from the back of the launch. By mid-May the divers had retrieved $123,995.

A week later catastrophe struck. Severe gales sent waves 30 metres up the cliffside. One stupendous roller snapped the derrick in two. Moments later it was reduced to matchwood. The crew saved the air pumps but the large diving bell was damaged beyond repair. Another storm wrecked the small bell. Then Moore and two sailors drowned when a wave swamped their boat. Undeterred, Dickinson built two new bells. And with no chance of replacing the derrick, he reverted to his original plan and fixed a cable across the cove.

Working from the cable was not as efficient but the divers continued to bring up buckets of bullion. In March 1832, more than a year after he began the operation, Dickinson was ordered back to Rio and Captain J. F. F. de Roos of the brig *Algerine* took over. Dickinson and his long-suffering crew had salvaged $588,801. With Dickinson's equipment, de Roos fetched up a further $161,500.

The salvage was a spectacular success and Dickinson was vindicated. All but a sixteenth of the treasure was recovered – along with guns, shot and copper and iron fittings from the ship. Dickinson found fame and fortune more elusive. The Court of Admiralty eventually awarded £29,000 salvage. Dickinson was allocated a quarter share – followed shortly by a bill. For wear and tear on HMS *Lightning* and food and wages for its crew, the Admiralty demanded £13,833 – almost twice Dickinson's share of the reward.

## ☀ Ragbag rainforest

*On top of Green Mountain, something strange has been happening. This remote mountain poking into the trade winds from the British imperial outpost of Ascension Island in the mid-Atlantic has created its own fully functioning cloud rainforest ecosystem virtually from scratch in just 150 years. It did it from odds and ends of botanical scrap brought in by the Royal Navy. According to ecological theory, rainforests are supposed to evolve slowly over millions of years, as species co-evolve and ecological niches are created and filled. Discovering the Green Mountain cloud forest is like finding that a pile of used car parts in a scrapyard has spontaneously reassembled into a functioning car. Unless, that is, ecologists have got their theories hopelessly wrong.*

When Charles Darwin stopped off at the mid-Atlantic island of Ascension in July 1836, homeward bound after his long journey aboard the *Beagle*, he described an island 'entirely destitute of trees'. The near-naked island was no casualty of human activity, however. Eighty years before, when Ascension was still uninhabited except for a few passing sailors, Swedish botanist Peter Osbeck dropped by on his way home from China. He wrote of 'a heap of ruinous rocks' with a bare, white mountain in the middle.

At about a million years old, the volcanic island was a geo-logical upstart and scarcely able to get going biologically because of its remoteness – some 2,000 kilometres from the nearest continent. All it boasted was a couple of dozen species of plant, most of them ferns and some of them found nowhere else.

And so it might have remained. But in 1843, British plant collector Joseph Hooker made a brief call on his return from Antarctica. Surveying the bare earth, he concluded that the island had suffered some ecological calamity that had denuded it of vegetation and triggered a decline in rainfall that was turning the place into a desert.

The Royal Navy, which by then maintained a garrison on the island, was keen to improve the place and asked Hooker's advice. He suggested an ambitious scheme for planting trees and shrubs that would revive rainfall and stimulate a wider ecological recovery. Perhaps lacking anything else to do, the sailors set to with a will.

In 1845, a naval transport ship from Argentina delivered a batch of seedlings. More than 200 species of plant arrived from the Cape Botanic Gardens in South Africa in 1858. In 1874, Kew sent 700 packets of seeds, including those of two species that especially liked the place: bamboo and prickly pear.

With sailors planting several thousand trees a year, the bare white mountain was soon cloaked in green – and renamed Green Mountain. An Admiralty report in 1865 praised the new forest growing in the clouds on top of Ascension. The island 'now possessed thickets of upwards of 40 kinds of trees besides numerous shrubs', it said. And it noted that 'through the spreading of vegetation, the water supply is now excellent'.

By the early 20th century the mountain's slopes were covered in guava, banana and wild ginger, the white-flowered *Clerodendrum* and Madagascan periwinkle, as well as the Norfolk Island pine and mighty eucalyptus from Australia. Up on the summit a bamboo forest, buffeted by the fierce trade winds, was soon howling like a giant wind chime.

Modern ecologists throw up their hands in horror at what they see as Hooker's environmental anarchy. The exotic species wrecked the indigenous ecosystem, squeezing out the island's endemic plants. In fact, Hooker knew well enough what might happen. 'The consequences to the native vegetation of the Peak will, I fear, be fatal, and especially to the rich carpet of ferns that clothed the top of the mountain when I visited it.' However, he and the navy saw greater benefit in improving rainfall and encouraging a more prolific vegetation on the island.

But there is a much deeper issue here than the relative benefits of sparse endemics versus lush aliens. And as botanist David Wilkinson of Liverpool John Moores University in the UK recently pointed out after visiting the island, it goes to the heart of some of the most dearly held tenets of ecology. Conservationists' understandable concern for the fate of Ascension's handful of unique species has, he says, blinded them to something quite astonishing – the fact that the introduced species have been a roaring success.

Today's Green Mountain, says Wilkinson, is 'a fully functioning man-made tropical cloud forest' that has grown from scratch from a ragbag of species collected more or less at random from all over the planet.

How could it have happened? Conventional ecological theory says that complex ecosystems such as cloud forests can emerge only through evolutionary processes in which each organism develops in concert with others to fill particular niches. Plants co-evolve with their pollinators and seed dispersers, while the microbes in the soil evolve to deal with the peculiarities of the biochemistry of the leaf litter.

But that's not what happened on Green Mountain. And the experience suggests that perhaps natural rainforests are con-structed far more by chance than by evolution. There is a term for this among dissident ecologists. They call it 'ecological fitting'. Species, they say, don't so much evolve to create ecosys-tems as make the best of what they have. What works works.

'The Green Mountain system is a spectacular example of ecological fitting,' says Wilkinson. 'It is a man-made system that has produced a tropical rainforest without any co-evolution between its constituent species.'

Not everyone agrees. Alan Gray, an ecologist at the Centre for Ecology and Hydrology in Edinburgh, argues that the surviving endemic species on Green Mountain, though small in number, will still be co-evolving and may form the framework of the new ecosystem. The incomers may just be an adornment with little structural importance for the ecosystem. Even the new species may not be quite such a random selection as at first appears. 'Many of the imports may have come from the same place, importing their co-evolutionary relationships,' he suggests.

Yet the idea of the instant formation of rainforests sounds increasingly plausible as research reveals that supposedly pristine tropical rainforests from the Amazon to south-east Asia may in places be little more than the overgrown gardens of past rainforest civilisations.

The most surprising thing of all is that no ecologists have thought to conduct proper research into this human-made rainforest ecosystem. A survey of the island's flora conducted in the late 1990s by the University of Edinburgh and a subsequent conservation strategy by the British government were concerned only with endemic species. They characterised everything else – the majority of the island's flora and fauna – as a threat. In 2005 the Ascension authorities designated Green Mountain a national park and earmarked introduced species for culling rather than conservation.

Conservationists, Wilkinson says, have understandable concerns. At least four endemic species have gone extinct since the exotics started arriving. And five endangered plants are just clinging on. But in their urgency to protect endemics, ecologists are missing out on the study of a great enigma.

'As you walk through the forest, you see lots of leaves that have had chunks taken out of them by various insects. There are caterpillars and beetles around,' says Wilkinson. 'But where did they come from?' Are they endemic or alien? If alien, did they come with the plant on which they feed or did they discover it on arrival? Such questions go to the heart of how rainforests happen.

The Green Mountain forest holds many secrets. And the irony is that the most artificial rainforest in the world could tell us more about rainforest ecology than any number of natural forests.

## ☀ Two men and a wheelbarrow

*In the summer of 1951, John Bolton and Bruce Slee were working at Australia's Dover Heights radio-physics field station, high on a cliff in Sydney's eastern suburbs. They were exploring the origins of the enigmatic radio waves reaching Earth from space, and to make any progress they needed to pinpoint the sources of those signals. For that, they needed a very large dish. In the post-war years money was tight and materials scarce, so they set about making one themselves from cast-off gear and other people's rubbish. For 3 months, they spent every lunch break secretly digging a huge hole in the ground. Shaped to form a dish and lined with metal ties from packing cases, the hole became the world's second-largest radio telescope. And it worked so well, it found the centre of the galaxy.*

Secrecy was vital. It was better that no one knew what John Bolton and Bruce Slee were doing in their lunch breaks. The two young scientists were working at a redundant wartime radar station perched on the cliffs just south of the entrance to Sydney harbour. Like many others who had worked on secret wartime radar research, Bolton and Slee had become accidental astronomers, putting their knowledge of radio waves to use in the embryonic science of radio astronomy.

By 1946, the old military blockhouse on the cliff top had sprouted antennas that looked to the heavens rather than out to sea. Bolton and Slee had been given the job of monitoring the newly discovered radio emissions from the sun. But the sun was going through a quiet phase and they were getting bored. So they turned their antennas away from the sun and hunted for other celestial objects that might be producing radio waves. Their search was short-lived. When their boss, Joe Pawsey, noticed the antennas were not pointing where they were supposed to, he confined them to the lab.

The following year, Pawsey relented and allowed Bolton and Slee, now joined by the young engineer Gordon Stanley, to resume their observations and by 1952 they had earned quite a reputation for detecting new radio sources. They were still using primitive Yagi antennas – simple masts with crosspieces which they made from bits of old radar equipment. Despite this, they had picked up the big bursts of radio waves associated with solar flares and sunspots, showed that some of the new 'radio stars' were objects already familiar to optical astronomers, and made the first observations of galaxies beyond our own.

In an attempt to narrow down the positions of radio sources and produce a detailed map of the radio sky, they built increasingly complex antennas. By 1951, they had an array of 12 Yagi antennas scanning the skies, making it one of the world's most powerful radio telescopes. But Bolton wanted to investigate the structure of the objects they were finding and learn more about the processes that were generating the radio waves. For this they needed a more sensitive telescope that operated over a wide range of frequencies. 'We decided to build a dish: it would have better resolution and higher sensitivity,' says Slee.

But what sort of a dish? Money was tight and the young astronomers suspected Pawsey would veto the idea. They'd have to make their dish from whatever materials they could lay their hands on, and in secret – at least until it had proved its worth. Their inspiration came from a dish built a few years earlier by Bernard Lovell in England. It was designed to detect cosmic ray showers, but had also been used with some success for radio astronomy. Lovell's dish was 70 metres across, a size that was possible only because it was fixed to the ground. The Australians decided they would also make a fixed dish, but unlike Lovell's, which was built on a framework of posts, they would dig theirs out of the ground.

Just over the top of the cliff, out of sight of the blockhouse, was a sandy ledge. The digging would be easy and they wouldn't be seen. 'We knew we'd have to dig a big hole and that would involve a lot of labour,' recalls Slee. Each lunch break, he and Bolton slipped off with a couple of shovels and a wheelbarrow. It took 3 months to carve out a depression 21.9 metres across. To make it parabolic they moved sand from the centre to the edges to build a low rim, and shaped the interior with a wooden template that rotated around a central pivot. 'We shifted around 1,500 cubic metres of sand – that's an awful lot of barrow loads,' says Slee.

Once the hole was roughly the right shape, they packed down the sand and consolidated it with ash to make as smooth a surface as they could. Getting hold of the ash was no problem: every few days, Stanley, who was in on the plot, scrounged a truckload of the stuff from the nearby power station. 'They had to dump it somewhere so they were quite pleased that someone would take it away,' says Slee.

The dish now needed a reflective surface to focus incoming radio waves onto a small dipole antenna at the top of a central mast. This time they headed to the dockside at Botany Bay and scavenged cast-off steel ties from packing cases, which they laid at 30-centimetre intervals inside the dish. They now had the world's second-largest dish – albeit one that was rooted to the spot and able to monitor only the narrow strip of sky that passed overhead. It was time to see what it could do.

They began with a quick survey of the central region of

the Milky Way at a frequency of 160 mhz. Radio astronomers were keenly interested in this part of the heavens, because somewhere out there, shrouded in cosmic dust and gas, lay the very centre of our galaxy. The Australians were ideally placed to search for it because the region around the galactic centre passes directly over Sydney.

Although the crudity of their reflecting surface restricted them to long wavelengths, they produced a map with a resolution three times as fine as they could manage with a Yagi array.

It was time to confess to Pawsey. They had expected a reprimand, but Pawsey was so delighted he suggested they might treble the resolution again if they expanded the dish and gave it a better surface, allowing it to operate at a much higher frequency. No longer limited to using leftovers, Bolton and Slee increased the dish's diameter to 24.4 metres, building the new, higher rim on a scaffold of aluminium tubes and tensioned wires. They poured concrete over the sand-and-ash surface and replaced the steel strips with an all-over covering of chicken wire.

Newcomer Dick McGee, sometimes assisted by Slee, began to repeat the earlier survey of the central Milky Way at the new operating frequency of 400 megahertz. Now the features producing radio emissions began to come into sharper focus, and individual sources stood out more clearly. Most importantly, the astronomers could make out the small but powerful source known as Sagittarius A. Bolton had little doubt that this was the nucleus of the galaxy, the centre around which all the stars rotate. Optical telescopes had never been able to penetrate the cosmic dust to pinpoint the spot, but the hole-in-the-ground dish had done it.

In 1958, the International Astronomical Union adopted the spot as zero longitude and latitude in the galactic system of coordinates. By then, the Dover Heights station had been abandoned. The local council filled in the hole, turfed over the site and created a public playing field known as Rodney

Reserve. There, a few metres from the northern goalpost of a football pitch, the hole-in-the-ground telescope lies just beneath the grass.

## ☀The ant and the mandarin

*In 1476, the farmers of Berne in Switzerland decided there was only one way to rid their fields of the cutworms attacking their crops. They took the pests to court. The worms were tried, found guilty and excommunicated by the archbishop. In China, farmers took a more practical approach to pest control. Rather than rely on divine intervention, they put their faith in frogs, ducks and ants. Frogs and ducks were encouraged to snap up pests in the paddies and the occasional plague of locusts. But the notion of biological control began with an ant. More specifically, it started with the predatory yellow citrus ant* Oecophylla smaragdina, *which has been polishing off pests in the orange groves of southern China for at least 1,700 years.*

For an insect that bites, the yellow citrus ant is remarkably popular. Even by ant standards, *Oecophylla smaragdina* is a fearsome predator. It's big, runs fast and has a powerful nip – painful to humans but lethal to many of the insects that plague the orange groves of Guandong and Guangxi in southern China. And for at least 17 centuries, Chinese orange growers have harnessed these six-legged killing machines to keep their fruit groves healthy and productive.

Citrus fruits evolved in the Far East and the Chinese discovered the delights of their flesh early on. As the ancestral home of oranges, lemons and pomelos, China also has the greatest diversity of citrus pests. And the trees that produce the sweetest fruits, the mandarins – or kan – attract a host of plant-eating insects, from black ants and sap-sucking mealy bugs to leaf-devouring caterpillars. With so many enemies, fruit growers clearly had to have some way of protecting their orchards.

The West did not discover the Chinese orange growers' secret weapon until the early 20th century. At the time, Florida was suffering an epidemic of citrus canker and in 1915 Walter Swingle, a plant physiologist working for the US Department of Agriculture, was sent to China in search of varieties of orange that were resistant to the disease. Swingle spent some time studying the citrus orchards around Guangzhou, and there he came across the story of the cultivated ant. These ants, he was told, were 'grown' by the people of a small village nearby who sold them to the orange growers by the nestful.

The earliest report of citrus ants at work among the orange trees appears in a book on tropical and subtropical botany written by Hsi Han in AD 304. 'The people of Chiao-Chih sell in their markets ants in bags of rush matting. The nests are like silk. The bags are all attached to twigs and leaves which, with the ants inside the nests, are for sale. The ants are reddish-yellow in colour, bigger than ordinary ants. In the south if the kan trees do not have this kind of ant, the fruits will all be damaged by many harmful insects, and not a single fruit will be perfect.'

Initially, farmers relied on nests which they collected from the wild or bought in the market – where trade in nests was brisk. 'It is said that in the south orange trees which are free of ants will have wormy fruits. Therefore the people race to buy nests for their orange trees,' wrote Liu Hsun in *Strange Things Noted in the South*, written about AD 890.

The business quickly became more sophisticated. From the 10th century, country people began to trap ants in artificial nests baited with fat. 'Fruit-growing families buy these ants from vendors who make a business of collecting and selling such creatures,' wrote Chuang Chi-Yu in 1130. 'They trap them by filling hogs' or sheep's bladders with fat and placing them with the cavities open next to the ants' nests. They wait until the ants have migrated into the bladders and take them away.' Farmers attached the bladders to their trees, and in time the ants spread to other trees and built new nests.

By the 17th century, growers were building bamboo walkways between their trees to speed the colonisation of their orchards. The ants ran along these narrow bridges from one tree to another and established nests 'by the hundreds of thousands'.

Did it work? The orange growers clearly thought so. One authority, Chhii Ta-Chun, writing in 1700, stressed how important it was to keep the fruit trees free of insect pests, especially caterpillars. 'It is essential to eliminate them so that the trees are not injured. But hand labour is not nearly as efficient as ant power...'

Swingle was just as impressed. Yet despite his reports, many western biologists were sceptical. In the West, the idea of using one insect to destroy another was new and highly controversial. The first breakthrough had come in 1888, when the infant orange industry in California had been saved from extinction by the Australian vedalia beetle. This beetle was the only thing that had made any inroads into the explosion of cottony cushion scale that was threatening to destroy the state's citrus crops. But, as Swingle now knew, California's 'first' was nothing of the sort. The Chinese had been experts in biocontrol for many centuries.

The long tradition of ants in the Chinese orchards only began to waver in the 1950s and 1960s with the introduction of powerful organic insecticides. Although most fruit growers switched to chemicals, a few hung onto their ants. Those who abandoned ants in favour of chemicals quickly became disillusioned. As costs soared and pests began to develop resistance to the chemicals, growers began to revive the old ant patrols. They had good reason to have faith in their insect workforce.

Research in the early 1960s showed that as long as there were enough ants in the trees, they did an excellent job of dispatching some pests – mainly the larger insects – and had modest success against others. Trees with yellow ants produced almost 20 per cent more healthy leaves than those without. More recent trials have shown that these trees yield just as big a crop as those protected by expensive chemical sprays.

One apparent drawback of using ants – and one of the main reasons for the early scepticism by western scientists – was that citrus ants do nothing to control mealybugs, waxy-coated scale insects which can do considerable damage to fruit trees. In fact, the ants protect mealy bugs in exchange for the sweet honeydew they secrete. The orange growers always denied this was a problem but western scientists thought they knew better.

Research in the 1980s suggests that the growers were right all along. Where mealy bugs proliferate under the ants' protection they are usually heavily parasitised and this limits the harm they can do.

Orange growers who rely on carnivorous ants rather than poisonous chemicals maintain a better balance of species in their orchards. While the ants deal with the bigger insect pests, other predatory species keep down the numbers of smaller pests such as scale insects and aphids. In the long run, ants do a lot less damage than chemicals – and they're certainly more effective than excommunication.

# 10   Sleuthing with science

And finally … history may tell us much about the progress of science, but science can also tell us much about history. Here, scientists turn detective to unravel some very peculiar puzzles from the past.

## ☀ Riddle of the fronds

*Each year, on 14 April, Japanese seaweed farmers gather on a headland overlooking the Ariake Sea, a deep notch carved out of the southern island of Kyushu. Many of the farmers make their living in the waters below, others travel from coastal towns and villages elsewhere in Japan. They come to celebrate the life of a woman who lived and worked in Manchester, in the industrial north of England. Kathleen Drew had no connection with Japan, except for one thing – an enthusiasm for seaweed, and in particular a silky, reddish weed called* Porphyra. *In Japan, people wrapped rice in it and ate it. In Manchester, Drew studied it. She would never know that people in a country she had never visited knew her name and held a festival in her honour. But when she discovered* Porphyra's *best-kept secret, she saved many Japanese families from destitution and sowed the seeds of a great industry.*

In the little villages along the shores of the Ariake Sea, fishermen had harvested the seaweed they called nori for longer than anyone could remember. Dried into paper-thin sheets,

the glistening alga was a traditional delicacy. Demand out-
stripped supply, so the fishermen could sell all they collected.
But nori was one of nature's more unpredictable harvests.
The first delicate ribbons appeared around October – as if
from nowhere. By December, the silky fronds were ripe for
picking. By spring, the fronds began to wither away, and by
summer the seaweed had vanished. Some years it didn't
appear at all. In the good years, the villagers fished in summer,
gathered nori in winter, and flourished. When the nori failed,
they struggled to feed their families.

Nori has been a part of Japanese culture for at least 1,500
years. By the 8th century it was a prized dish, but one that
was so rare and costly that only the noblest households could
afford it. Then, in the 17th century, a few enterprising
fishermen began to cultivate the weed in sheltered bays and
inlets. Their methods were crude but they worked.

Certain spots around the coast were known as good nori
places, where the weed sprouted from the rocks, from jetties
and pilings – in fact, from any submerged surface in the shal-
lows. In early autumn, nori growers would take some bamboo
sticks and ram them into the mud close to the shore. A few
days later, they pulled up their poles, replanted them in water
near their own villages, and waited. Sure enough, the poles
were soon festooned with nori. By winter, the crop was ready
to collect and spread out to dry. Cultivation increased output
in the good years, but it didn't make the harvest any more
certain. Some years the weed still failed to appear.

By the early 20th century, marine biologists in Japan were
desperately trying to solve the great nori mystery. Where did
the seaweed go in summer? And why did it sometimes fail to
reappear? If they could fill in the missing months in the life of
*Porphyra*, they might be able to remove the uncertainty from
nori cultivation.

*Porphyra* is a member of one of the most ancient groups of
plants, the rhodophytes or red algae – organisms with a
complicated life history. The alga grows from spores that

settle onto any firm surface in shallow water, including bamboo poles. But no one knew where those spores came from. Although mature nori fronds produce spores of their own, they are not the same, and like the fronds themselves, they would vanish in the summer.

In 1948, disaster struck Kyushu. Typhoons destroyed many fishing boats. The pearl oysters were killed by chemicals draining into the sea from the paddy fields. And the nori crop failed. With no boats, no oysters and no prospect of any nori, many of the men from the fishing villages headed off to the island's coal mines to look for work. They had no reason to suspect their fortunes were about to change.

Halfway round the world, in a lab at the University of Manchester, botanist Kathleen Drew was wrapping up her study of purple laver, another species of *Porphyra*. Laver grows on rocky coasts around the British Isles and, like nori, it is edible. People living along the shores of the Bristol Channel – particularly in South Wales – eat the seaweed, not crisp and dried like the Japanese but boiled to a blackish gloop known as laver bread. Like nori, laver disappeared in summer and reappeared a few months later. After nine years of detective work, Drew had discovered where it went.

As with its Japanese counterpart, the purple fronds of laver produced spores, but these did not sprout new laver plants. Drew was as mystified as everyone else. Year after year, she collected laver spores and tried to persuade them to germinate. Her husband, Henry Wright Baker, professor of engineering at the university, built tidal tanks for her in which she could control temperature and light, and mimic the tides around the Welsh coast. But nothing she did would persuade the spores to sprout.

Drew decided to try a different tactic. Perhaps the spores only sprouted if they settled on something hard, where an emerging plantlet could get a firm roothold. Drew filled a flask with seawater, added some laver spores, and dropped in a carefully sterilised piece of oyster shell. Then the strangest thing happened. Out of the spores emerged not tiny leafy laver germlings but slender crimson threads. The threads began to burrow their way into the shell, and once inside grew into a network of fine red filaments. The same thing happened whether she threw in an oyster, an old cockle shell or even a piece of discarded eggshell. After a few months the red threads turned, and their tips began to emerge from the surface of the shell. There they sprouted little pink tufts that produced spores of their own.

This organism was nothing like laver. In fact, marine botanists knew it as another species entirely, one they called *Conchocelis rosea*. Drew now knew that *Conchocelis* was just a phase in the laver's life. And if that was so, then the 'conchospores' escaping from the tiny pink tufts should grow into new leafy laver plants. They did.

In 1949, Drew's revelation appeared in the journal *Nature*. At last there was an explanation for the years when laver, and nori, went missing. Stormy weather interrupted the supply of conchospores, either by flinging *Conchocelis*-infested shells high up the shore or dragging them far out to sea.

The news soon reached Japan. At Kyushu University in Fukuoka, marine botanist Sokichi Segawa was astonished. It seemed so improbable. But if Drew was right, and all *Porphyra* were like laver, then it might be possible to culture the spores artificially and so guarantee a good nori harvest. First he repeated Drew's experiments with nori: the result was the same. Then he went to see a man who could put the discovery to good use.

Fusao Ota, a marine biologist at the local fisheries research laboratory at Kumamoto, knew well the difficulties and hardships the nori farmers faced. When they were in trouble, they knocked on his door. Yet there was little he could do for them. Now, perhaps, he could change that. By 1953, Ota had perfected a method of artificially seeding nets with nori, a breakthrough that paved the way for the mass production of today.

Ota's method was simple but effective. He stirred mature nori fronds in a vat of seawater so that they released their spores, then tipped the suspension into tanks containing oyster shells – leftovers from the cultivated-pearl industry. When the shells sprouted pink tufts, he stirred the water again to shake out the conchospores. Into this mixture he dangled ropes and then strung them between poles in the shallow water near the lab. Nori fronds quickly sprang from the ropes. The nori growers were back in business.

Today, nori cultivation is the world's most profitable form of aquaculture, a business worth a billion dollars a year. Growers work in collectives, taking their nets to be 'seeded' at government-run centres where spores are cultured in huge tanks, safe from the vagaries of nature. At the end of each season, the growers return with some of their finest fronds to provide spores for the next season.

Drew never found out that her botanical detective work had saved Japan's seaweed industry: she died in 1957 at the age of 56. But the nori growers knew how much they owed her. On 14 April 1963, they unveiled a polished granite memorial bearing the likeness of the woman they called the Mother of the Sea. And they have been back to pay their respects – and bring offerings of seaweed – every year since.

## Death and the outcast

*In the summer of 2000, robbers ransacked an ancient family burial cave at Akeldama in the Hinnom Valley in Jerusalem. Thieves had been there before, smashing open the stone caskets that held the bones of the dead. This time the looters found a lower burial chamber, and in the depths discovered a secret sepulchre, a niche in the rock that had been blocked with a boulder and sealed with cement. By the time archaeologist Shimon Gibson arrived a few hours later, the thieves had done their worst and fled. But they had left something.*

*Inside the sepulchre lay a tangled mess of brown fibres. These were the remains of a man still wrapped in his burial shroud. When this man died, it was customary to lay the dead in a niche and return later to rebury the bones in a stone chest. Why had no one come back for this man's bones? When archaeologists discovered the answer, they also found the key to a much greater puzzle.*

When tomb robbers found an intact sepulchre in a tomb at Akeldama, they immediately set to work to free the stone that blocked the entrance. The stone had been sealed in place with strong cement-like mortar and it took considerable effort to hack it out. With the stone gone, one thief knelt down to look inside. 'You could see the imprint of his knees in the soil,' says Shimon Gibson, a member of the archaeological rescue team that rushed to the scene.

The grave was little more than a narrow tunnel in the rock, and even with a torch it was hard to make out what the unprepossessing brown heap inside might be. 'They were after caskets or jewels – something valuable. When they saw the brown mess, they left it.'

Closer inspection revealed that the brown material was a mixture of hair and the remnants of a wool-and-linen burial shroud. There were fragments of bone too. Carbon dating of the shroud put the man's death somewhere in the first half of the 1st century AD. And the man clearly came from a family of note. 'The tomb was at the foot of Mount Zion, a stone's throw from the city and in a priestly and aristocratic quarter,' says Gibson. A neighbouring tomb belonged to the family of Caiaphas, the high priest who delivered Jesus into the hands of Pontius Pilate.

Such a man might expect to receive the customary burial rites: but instead of leaving his body in a niche for a year or so and then reburying his bones in an ossuary, his family had walled him up in the rock. 'Every other skeleton was in an ossuary,' says London-based palaeopathologist Mark Spigelman. 'This was a high-class, deeply religious family.

There has to be a reason why they didn't observe the normal rituals. Either there was some sort of social upheaval that prevented it, in which case we would never know, or they were too scared to do it.' The fact that someone sealed the niche so carefully suggests they never intended to open it again. 'That made me think they were afraid of something,' says Spigelman. 'And that made me think maybe he was a leper.'

Spigelman and his colleague Helen Donoghue, an expert on ancient DNA at University College London, decided to test his hunch. As part of the team investigating the tomb, they had been asked to look for DNA from *Mycobacterium tuberculosis*, the bacterium that causes TB. Some bones from the tomb showed the hallmarks of infection, but only the presence of bacterial DNA would confirm the diagnosis. They found plenty. The shrouded man was so heavily infected he probably died from the disease. But samples of his bone produced DNA from a second mycobacterium, *M. leprae*. As Spigelman suspected, the man had leprosy.

In China and India leprosy is an ancient disease, but there is a big question mark over when it reached the Middle East. The 'lepers' of the Bible's Old Testament were not suffering from leprosy, says Gibson. The word referred to people with skin diseases in general and to those regarded as spiritually unclean. But did this apply to the lepers of the New Testament? 'Now that we have a confirmed case from Jerusalem in the 1st century, it's more likely that the lepers of the New Testament did genuinely have the disease,' says Gibson.

For Donoghue and Spigelman the discovery was significant for a different reason. They were intrigued by the fact that the shrouded man had widespread and active infections of both TB and leprosy. For almost a century, historians and archaeologists have accepted the theory that infection by *M. tuberculosis* makes the body immune to *M. leprae*, a notion that neatly explained why leprosy suddenly began to disappear from western Europe in the late Middle

Ages. As TB spread, encouraged by migration from the land into towns, leprosy began to retreat. TB kills only a fraction of those it infects, and around 90 per cent show no signs of disease. So as TB became endemic it created a huge pool of people immune to leprosy.

If leprosy was a recent arrival in the Middle East, it came even later to western Europe, carried north and west by the Romans. Later still, the Vikings took it home to Scandinavia, while the Crusaders probably spread it about still more. By the 12th century the disease had reached a peak. Leprosy is hard to transmit and rarely fatal, yet it was feared like no other disease. Those diagnosed with it were immediately cast out from society and subject to harsh laws. They lost family, property and livelihoods, and relied on charity for survival. But by the 13th century the disease was in sharp decline, and by the 16th century it was virtually gone. In its place was the new scourge of tuberculosis.

The discovery that the man from Akeldama had both TB and leprosy prompted Donoghue and Spigelman to re-examine samples from other ancient sites. They found DNA from both bacteria in bones from Dakhleh Oasis, a 4th-century Egyptian shrine visited by people with leprosy, at Püspökladàny, a 10th-century burial ground in Hungary, and at a Viking-age cemetery in northern Sweden. 'These diseases clearly coexisted in the past,' says Donoghue. The idea of cross-immunity was beginning to look a bit shaky.

The theory arose because *M. tuberculosis* and *M. leprae* have some identical antigens, which trigger production of identical antibodies. Even a symptomless TB infection would leave the body equipped with antibodies against invading *M. leprae*. But antibodies are not the body's main line of defence against mycobacteria. These pathogens operate from inside their host's cells, and it takes bigger guns – the killer cells and macrophages of the cell-mediated immune response – to defeat them.

Donoghue and Spigelman think leprosy declined not

because TB protected against it, but the opposite: leprosy made people extremely vulnerable to TB. Their defences were already down, making them easy prey for the big killer. People who develop the most disfiguring form of leprosy are known to have a defect in their cell-mediated immune response. But even without such a breach in their defences, people who had leprosy in the Middle Ages would have had few resources to fight a second infection. 'These people suffered terrible social stigma,' says Donoghue. 'They were isolated and moved on. They relied on alms and probably never had enough to eat. And they would have been depressed. All these things would have had a profound effect on their immune response. They would be very susceptible to TB.' Anyone with a latent TB infection who went on to contract leprosy would not live long either. 'If they got leprosy, the stress and stigma would lead to immunosuppression and the TB would emerge and kill them.' TB, then, made short work of Europe's outcasts until there were too few to maintain the disease.

The man in the shroud was no outcast. His family abandoned him only after he was dead. 'He didn't have to beg for alms or scrape a living among the rubbish dumps and sewers of Jerusalem,' says Gibson. 'We found his hair was clean and well kept. He was cared for and looked after.' But even so, leprosy eventually wore down his immune system and made him prey to TB.

## ☀Sleepwalking in Springfield

*In July 1833, Jane Rider, servant to the Stebbins family of Spring-field, Massachusetts, rose from her bed, dressed and made her way downstairs. Quickly and efficiently, she set the table for breakfast, arranging the crocks, cutlery and condiments with the utmost pre-cision. She negotiated a narrow doorway with a large tray laden with coffee cups then went to the pantry to skim the milk. Without spilling a drop, Rider poured the cream into one cup, the milk into another. She cut the bread, put the slices on a plate and divided them neatly down the middle. None of this would be at all strange except that it was the middle of the night and Jane had her eyes shut tight.*

Festus Stebbins was alarmed. The family's new servant had started to do strange things in the night. The girl, a fresh-faced 16-year-old, joined the household in April 1833. At the time, she seemed healthy and well suited to her new place. Her father was 'an ingenious and respectable mechanic' in Brattleboro, Vermont, and Rider was intelligent, obliging and 'better educated than most in her position'. Stebbins had no reason to suspect she would disrupt the family's sleep and attract a stream of visitors to his home.

Rider had her first 'paroxysm' on 24 June 1833. As she fought to get out of bed, the family fought to keep her in it. Her face was flushed, her pulse raced and she complained of an intense pain in the side of her head. Stebbins thought she was deranged and sent for Lemuel Belden, a local doctor with an interest in the workings of the mind. He blamed the attack on undigested food in her stomach and gave her an emetic. Rider went back to sleep. The next day she remembered nothing.

A month later, Rider suffered a second paroxysm. This time the family gave up the struggle to keep her in bed and watched what happened next. Rider dressed, went to the kitchen and began to prepare breakfast. 'She went through the whole operation ... with as much precision as in the day with her eyes closed and no light,' Belden reported. Duties done, Rider returned to bed. Next morning, when she discovered the breakfast laid, she asked why she had been allowed to sleep on while someone else did her work.

In the early 19th century, there was intense interest in somnambulism. Were sleepwalkers mad or somehow guided by spirits? Public fascination with these notions prompted a rash of popular songs, plays and books featuring sleepwalkers. In 1829, Bellini's opera *La Sonnambula* thrilled audiences with a heroine who wandered the rooftops in her sleep. Belden's ideas were more down to earth: he believed all mental processes could be explained by physiology. He thought sleepwalking was a state somewhere between dreaming and madness that was somehow triggered by physical disease. That being the case, it could be cured.

Rider's attacks grew more frequent. Sometimes she pottered about her room, tidying things away in drawers, which she was unable to find when she woke. Sometimes she sat bolt upright in bed and sang songs or recited poems she later denied she knew. Occasionally she performed more complicated tasks. She threaded a needle and sewed. She cooked dinner, fetching in the wood to light a fire, collecting vegetables from the cellar, then preparing and cooking each type in the right manner, testing them at intervals to see if they were ready.

What intrigued Belden most was that Rider appeared to acquire extraordinary visual skills during her attacks. The Stebbinses told him they were sure she could see in the dark and even with her eyes closed because she never groped for things and neatly sidestepped obstacles they put in her way. Belden had read reports of sleepwalkers who possessed amazing powers of perception or heightened intellectual abilities. Presented with a case of his own, he decided to investigate. On 10 November, Belden sat the sleeping servant in a corner where it was too dark to see and asked her to read from cards. No problem. He gave her coins and she called out their dates without hesitation.

By this time, the story of the Stebbinses' servant had made the pages of the *Springfield Republican* and curious townsfolk began calling at the house in the hope of seeing the Springfield somnambulist in action. As Rider's fame spread, their numbers increased. So did the attacks, which also began to happen during the day. Naturally enough, Belden wondered if Rider was faking. Yet he saw 'nothing in her character to suggest she was an impostor' and declared that 'anyone who witnesses her during a paroxysm is convinced nothing is feigned' because of the 'artlessness and consistency of her conduct'.

He needed to be sure, however, and so he arranged an experiment in front of witnesses. On 20 November, Belden placed cotton wads over Rider's eyes, then tied a black silk handkerchief lined with cotton around them. In the presence of a group of 'respectable, intelligent men from Springfield', he presented Rider with cards on which people's names were written. 'She read them as soon as they were presented to her and always held the paper right side up and brought it into the line of vision.' He handed her a watch. She opened the case and told him the time. Everyone there was satisfied there had been no deception. 'She seemed to see as well with her eyes closed as when open,' wrote Belden.

Fascinated though he was, Belden believed the aggravation caused by so many visitors would hamper Rider's recovery. If her disease was to be cured, she needed seclusion. And so Belden arranged for Rider to go to the newly opened Worcester Hospital for the Insane. There she was treated with powerful drugs, acid footbaths and a generous application of leeches to her head. Still convinced poor digestion was the root of her trouble, Belden insisted Rider was kept on a light diet and well away from fruit.

Within two weeks Rider had lost her apparent ability to see with her eyes closed. Her paroxysms grew less frequent and although not wholly cured of sleepwalking, after five months she went home to Brattleboro. While she faded into obscurity her case was made famous by Belden, who that year published his account in *Somnambulism: The extraordinary case of Jane C. Rider, the Springfield somnambulist*.

How did Belden explain Rider's behaviour? He had little trouble with the sleepwalking part. Illness, probably some digestive disorder, disturbed her sleep and prompted her wanderings. Her heightened night vision was more of a puzzle. He concluded that despite the darkness and blindfolds, 'when Jane read, wrote etc, she actually saw'. Ruling out miracles, he reasoned that during a paroxysm, Rider's eyes were peculiarly sensitive to light. He also argued that darkness is never absolute: some light must pass through even blindfolds and eyelids. To make sense of the faint images on her retinas, the part of Rider's brain that dealt with perception must also have been 'excited' to a higher level than normal.

A simpler explanation is that Rider suffered from not one, but two distinct sleep disorders – sleepwalking and a fugue or amnesic state – and Belden was right to think her case extraordinary. Much of her nocturnal behaviour was probably quite genuine, says Jim Horne, director of the Sleep Research Centre at Loughborough University in the UK. Sleepwalkers are in a deep sleep and move like automatons, unaware of their surroundings. 'When Jane stayed in her room and tidied things away in drawers, she was probably sleepwalking,' says Horne. 'But when she did more complicated tasks such as threading a needle or preparing dinner, she was in what's known as an amnesic state – wide awake and able to see but unable to remember it later.' Neither state is likely to have been triggered by eating too much fruit. 'She seems to have been a rather disturbed individual,' says Horne. 'Amnesic episodes are generally in troubled people and triggered by anxiety.'

And what of those incredible visual powers? 'That can only have been fraudulent,' says Horne. 'If your eyes are closed, you can't read. End of story.' Belden may have embellished the case to impress his colleagues. More likely he was tricked by a smart girl with a good memory who enjoyed being the centre of attention and learned how to pull off some fairly simple tricks, with or without an accomplice. That part of the case is unlikely ever to be solved.

# ☀ Dead man working

*'A fox, a wildcat and a dog go through a customs-post. They are taxed 111 cash. The dog says to the wildcat, and the wildcat says to the fox, "Your skin is worth twice mine; you should pay twice as much tax!"' Who pays what? In 186 BC, a civil servant employed by the Emperor of China was buried in a tomb in what is now Hubei Province, along with a few books he might want to read in the after-life. Among them was China's earliest known work on mathematics, a collection of problems, including the one about the fox, complete with answers and methods of calculation. Against expectations, not all the maths was practical, its purpose to keep the empire running smoothly. It seems even then some people did maths just for fun.*

In December 1983, archaeologists opened up one of the many tombs at an ancient burial ground near Zhangjiashan, in China's Hubei Province. The anonymous man in tomb 247 had been buried around 186 BC. He was neither rich nor famous: he was a civil servant, one of a great army of officials employed to run the new Chinese empire. His career started under Qin Shi Huang Di, 'The First Sovereign Lord of Qin', who in 221 BC destroyed all his rivals and established a unified empire that extended over much of what is now modern China.

When the First Emperor died in 210 BC, he was buried in a vast mausoleum along with his now famous army of terracotta warriors. Although the Qin dynasty didn't last long, the bureaucracy he set up to administer the state lived on, and the man destined for tomb 247 continued his official duties under the new Han dynasty. When he died, he too was buried with the things he thought he would need most in the next world: his books. One of them, the *Suan shu shu – Writings on reckoning* – is the oldest work on mathematics known from China.

Ancient Chinese books were written in ink on strips of

bamboo strung together to form a scroll. Unfortunately for the archaeologists, the strings had long since rotted, so instead of a neat stack of scrolls they found a jumble of more than 1,200 bamboo strips. Painstaking reconstruction produced a set of government statutes, law reports and writings on therapeutic gymnastics. Most exciting of all were the 190 strips that made up the *Suan shu shu*. It took Chinese scholars 17 years to fit them together in a way that made sense.

Once that was done, Christopher Cullen, director of the Needham Research Institute in Cambridge, UK, and a historian of Chinese mathematics, spent six years translating and studying the *Suan shu shu*. It is not a book in the usual sense, he says, but a compilation of problems put together by a mathematical magpie. Someone, perhaps the dead civil servant himself, had collected 69 problems from a variety of sources and stitched them into a sort of album. Each problem has a question, an answer and a method for reaching it, but the varied styles of working suggest they were written by a number of people. At least two – Mr Wang and Mr Yang – were proud enough of their creations to put their names to them.

The *Suan shu shu* extends what we know about the development of Chinese maths by three centuries. It has also produced surprises. Chinese mathematics has often been stereotyped by westerners as relentlessly practical, developed as a tool in the service of those who ran the empire, yet the *Suan shu shu* suggests that the early imperial age was also a time of mathematical creativity.

Mathematics in China developed along different lines from that in early western civilisations. The West's most influential ancient mathematician was Euclid, a Greek who lived in the 4th century BC. His approach, as laid out in his treatise on geometry, *Elements*, was deductive: he began with a small number of apparently indisputable truths, such as 'the whole is greater than the part', and on these foundations proved many less obvious things by logical deduction. As a result, the western tradition of maths is based on theorems and proofs.

While the Greeks dazzled with their abstract concepts and elegant proofs, the Chinese approach was fundamentally different. Chinese mathematicians began by providing workaday tools for people who needed to perform a raft of calculations in the highly organised new society they lived in. These generalised methods – algorithms – could be applied widely to many problems, and the aim was to generalise them still further to solve even more types of problem. 'The Chinese were the first people to talk explicitly about how to generate "software" that can do a lot of jobs,' says Cullen.

Before the opening of tomb 247, the earliest known Chinese maths text was the *Nine Chapters on Mathematical Procedures*, which dates from around AD 100. Like *Elements*, this was a treatise on maths that became well known and was in continuous circulation for many centuries. 'We've always wondered how Chinese maths got started,' says Cullen. 'Just as Euclid must have built on the work of earlier scholars, the material in the *Nine Chapters* must have come from somewhere. The *Suan shu shu* is as big a discovery as finding a book by one of Euclid's predecessors.'

Some of the maths in the *Nine Chapters* was in use early in the imperial age: by 186 BC, the distinctive Chinese way of doing things was already established. The scroll includes problems ranging from simple calculations with fractions to methods for working out the volumes of solid shapes. There are techniques for calculating tax rates, the productivity of workers and much else of use to the budding bureaucrat. What is not in the collection is also interesting. There is no method for extracting square roots, although there is a technique for getting good approximations, and the Chinese equivalents of simultaneous equations and Pythagoras's theorem – both in the *Nine Chapters* – are missing.

Clearly, there were people who had to perform a range of

calculations of varying degrees of difficulty, and there were
people who could provide methods that enabled them to do
the job efficiently. So was the *Suan shu shu* a handbook of
reckoning that enabled the man in tomb 247 to fulfil his daily
duties? 'Officials were expected to be able to do accounts.
They had to add and subtract, multiply and divide. But this
goes way beyond that,' says Cullen. 'Without being a Qin
official it's impossible to know whether some of the more
complex problems reflected any practical needs in the course
of one's work. But some of the problems are clearly designed
to make the reader rack their brains.' At least two have
absolutely no practical use, he says. 'They are intended to
show off the author's mathematical virtuosity.' So early
Chinese maths wasn't entirely mundane and useful. 'Right
from the beginning there were people interested in maths for
its own sake.'

To a keen mathematician like Cullen, that's no big
surprise. What he did find surprising was how people learned
about maths. In the early days of empire there was no well-
ordered textbook like the *Nine Chapters* to refer to. People
acquired knowledge piecemeal. When a mathematician came
up with a new method, he would write it down on a strip or
two of bamboo, and such snippets became the currency of
learning. When you heard about a new method, you could
ask for a copy and add it to your collection. 'It was a bit like
surfing the net and downloading bits you found interesting
into a file,' says Cullen.

So why did our anonymous official need his books after
he was dead? In pre-Buddhist China, people expected to go
to a shadowy place after death where they would carry on
much as before. 'His was a high-status job that made him a
person of importance. He would probably hope to continue
doing it in the next world and so he'd want to have his
equipment with him – in this case, his books.' And if he had
a little spare time after taxing and judging his shadowy peers,
he might relax with a little mathematics.

# ☀ Last days of the locust

*In the summer of 1873, black clouds drifted east from the foothills of the Rocky Mountains towards the newly settled farms of Nebraska, Iowa, Minnesota and the Dakotas. The pioneer families had no warning: the sky went dark at midday, the air filled with a sound like a thousand scissors. Then the clouds fragmented and locusts fell like hail onto crops of corn and wheat. In a few hours, the insects had devoured months of work. Locusts had been invading farms on the American frontier on and off for decades, but the irruption of the mid-1870s entered into legend. Many families gave up farming and fled to the cities. On 26 April 1877, John Pillsbury, the governor of Minnesota, called for a day of prayer to plead for deliverance from the locusts. A few days later, the insects rose up and left as inexplicably as they had come.*

When the Rocky Mountain locusts swarmed, they darkened the skies over vast swathes of the western and central USA, from Idaho to Arkansas. The number of insects was mind-boggling: one reliable eyewitness estimated that a swarm of locusts that passed over Plattsmouth, Nebraska in 1875 was almost 3,000 kilometres long and 180 kilometres wide. And they were devastating. 'You couldn't see that there had ever been a cornfield there,' one farmer said after a swarm passed through his land. Yet between these episodes of frantic fecundity, the locusts seemed to disappear.

The Rocky Mountain locust (*Melanoplus spretus*), a big, beefy species of grasshopper, was considered the greatest threat to agriculture in the West. So entomologists tried to learn everything they could about the insects – what triggered them to swarm, what they ate and how they reproduced. But after the spring of 1877, the locusts vanished and never plagued western farmers again. Within 30 years of Minnesota's official day of anti-locust prayer, the insect was extinct. The last live specimen was found by a river on the Canadian prairie in 1902.

No one mourned the loss, and scientific interest in the locust waned. In the 1940s and 1950s, when farmers began to wage war on their enemies with insecticidal chemicals, a few researchers began to speculate about what could possibly have seen off the locust so spectacularly in those pre-pesticide days.

During the disastrous outbreaks of the 1870s, farmers fought back with every tool they could find or invent. They deliberately set their fields on fire. They dragged tar-coated hunks of metal through the ground, hoping to trap locust hatchlings in the sticky goo. Nothing helped much. When desperate pioneer women tried to protect their vegetable gardens by draping blankets over them, the locusts ate the blankets before moving on to the vegetables.

Whatever had done for the locust, it seemed, was some event far beyond the capabilities of 19th-century farmers. As the extinction coincided with a time of dramatic environmental change across the West, there were plenty of plausible explanations. Perhaps the locusts had depended on the fires that Native Americans had routinely set to keep the prairies open. Or maybe their most crucial habitat had been shaped by the huge herds of bison that were now all but extinct.

Most standard entomology texts claimed that extreme fluctuations in population, like those that took place when the locusts swarmed, were a sign of a species in trouble, fighting to recover a balance with its environment. The sweeping changes that came with settlement, some scientists suggested, pushed the locusts through cycles of population explosion and collapse, and in the end wiped the species out.

When Jeffrey Lockwood, an insect ecologist at the University of Wyoming, was hired to explore the biology of grasshoppers in 1986, the post-mortem on the Rocky Mountain locust had not gone beyond this sort of general speculation. Lockwood wanted to know more, and he hoped that somewhere there were still a few specimens of the long-vanished locust to study. Among the high peaks of the

Rockies in Montana and Wyoming were glaciers where swarming insects had fallen, become immobilised by the cold, and died. Some of these glaciers might still hold frozen remains of the Rocky Mountain locust.

Lockwood and his students spent summers searching in the ice at remote spots high in the Rockies. They began their hunt at Grasshopper Glacier in Montana, hoping it might live up to its name. Sure enough, they found some scattered body parts that might have once belonged to Rocky Mountain locusts, but without whole bodies there was no way to prove these bits had not belonged to some other, still living, species of grasshopper.

'Finally, after four years of fruitless searching, we found the mother lode,' says Lockwood. On Knife Point Glacier in the Wind River Mountains of Wyoming, they recovered 130 intact bodies of Rocky Mountain locusts, the legacy of a swarm that had risen out of the river valleys of western Wyoming in the early 1600s. The antiquity of the frozen insects – confirmed by radiocarbon dating – proved that locusts had irrupted long before European settlers changed the face of the West. The reproductive frenzies, which at times produced enough insects to blanket the entire state of Colorado, were normal events in the history of the locust. Further study of Knife Point Glacier revealed deposits of locust parts throughout the layers of ice, indicating that swarms passed over the mountains at regular intervals during the centuries before the locust's extinction.

To find more clues to what killed off the locust, Lockwood began to scour the scientific literature of the late 1800s. There he found the writings of Charles Riley, an entomologist who had spent much of the 1870s and 1880s searching for ways to kill the locusts.

Riley had mapped what he called 'the permanent breeding zone' of the locust, the territory where mating adults and eggs could be found every summer, regardless of whether the locusts were swarming. For a species that could spread across much of the continent during outbreaks, this home base was surprisingly small. Between swarms, the locusts lived only in the river valleys of Montana and Wyoming, where they buried their eggs in the damp ground along the banks of streams. These fertile spots were the same places the incoming settlers chose to farm.

Riley had experimented with ways to control the number of locusts. Ploughing, he discovered, could push locust eggs so far down into the soil that they would fail to hatch. Flooding the ground where eggs had been deposited also killed many of the young. Riley concluded that agriculture itself – the processes of ploughing and irrigation – were the strongest weapons against the locust. But because less than 10 per cent of the land in the western USA was arable, he doubted that farming would ever have had a significant impact on the locust.

A century after the locust disappeared, Lockwood took Riley's map of locust egg-laying areas in the 'permanent zone' and superimposed it on a map of land under cultivation for corn, wheat or hay in 1880. He found that he had charted the geography of an extinction. In the 1880s, when the locust population had shrunk during an intermission between outbreaks, every corner of its breeding grounds was being farmed. The settlers, Lockwood suggests, had destroyed their nemesis without ever knowing it, simply by ploughing the land and watering their crops. 'The most spectacular "success" in the history of economic entomology – the only complete elimination of an agricultural pest species – was a complete accident,' he says.

# About the contributors

**Paul Collins** ('Scalpel, suture, salt beef'; 'Sunshine for sale'; 'Mark Twain's big mistake'; 'Live from the Paris Opera'; 'Henry's little pot of gold'; 'Hens' eggs and snail shells'; 'Nothing but a ray of light'; 'Arsonist by appointment'; 'Like a lead balloon'; 'As Shakespeare liked it') teaches creative writing at Portland State University, Oregon, and appears on National Public Radio as its 'literary detective'. His latest book is *The Murder of the Century*.

**Barry Fox** ('Their Führer's voice') wrote about patents and consumer electronics for *New Scientist* for over thirty years. He is now European contributing editor for *Consumer Electronics Daily*.

**Sharon Levy** ('Last days of the locust') is a contributing editor at *OnEarth*, and writes regularly for *BioScience*, *New Scientist*, and other magazines and websites. She is the author of *Once and Future Giants: what Ice Age extinctions tell us about the fate of Earth's largest animals*.

**Richard A. Lovett** ('White knuckles in Wyoming'; 'Surf's up') is a law professor turned science writer from Portland, Oregon. His articles have appeared not only in *New Scientist*, but also in *Nature*, *Science*, *National Geographic News*, *Cosmos*, *The Economist*, *Travel & Leisure* and *Psychology Today*. He also writes science fiction.

**Stephanie Pain** ('This won't hurt a bit'; 'The accidental aeronaut'; 'Inactive service'; 'A nose by any other name'; 'Dr Coley's famous fever'; 'The chunkiest chip'; 'Hats off to Mr Henley'; 'The great tooth robbery'; 'Lady of longitude'; 'Fruits of the tomb'; 'The impossible pound note'; 'A cure for curates'; 'The coachman's knee'; 'Into the mouth of hell'; 'Farmer Buckley's exploding trousers'; 'Night of the mosquito hunters'; 'The human centrifuge'; 'Pharaoh's ears'; 'Lamb's tales'; 'When men were gods'; 'War and peace'; 'Secret service'; 'Liquid assets'; 'Buckets of bullion'; 'Two men and a wheelbarrow'; 'The ant and the mandarin'; 'Riddle of the fronds'; 'Death and the outcast'; 'Sleepwalking in Springfield'; 'Dead man working') swapped life as an oceanographer in the deep Atlantic for a typewriter at *New Scientist*. She spent 15 years as a staff writer and editor, specialising in natural history and conservation, before creating and editing *New Scientist*'s popular Histories column. She is now a freelance writer based in Brighton.

**Fred Pearce** ('Land of the midnight sums'; 'Last-chance balloon'; 'Ragbag rainforest') is a freelance writer based in London. He has been environment consultant of *New Scientist* since 1992 and has reported for the magazine from 67 countries. His books, translated into 16 languages, include *When the Rivers Run Dry*, *Peoplequake* and *Confessions of an Eco Sinner*.

**Gail Vines** ('Toad in the hole'), former features editor at *New Scientist*, is now a freelance science writer who spends much of her spare time investigating the natural and cultural richness of a patch of Sussex downland east of Lewes, the source of Charles Dawson's flint.

**Geoff Watts** ('Turn on, tune in, stand back') is a journalist and broadcaster best known as the presenter of the BBC's long-running radio programmes *Medicine Now* and *Leading Edge*. He has written for many papers and magazines including the *Independent*, *Prospect*, *New Scientist*, the *Times Higher Educational Supplement* and the *Lancet*. He has written books on irritable bowel syndrome and the placebo effect.

# Acknowledgements

The editor would like to thank all those who helped make *New Scientist* 'Histories' happen, in particular the many museum curators, librarians, archivists and historians who shared their ideas and expertise. Tim Boon and his colleagues at the Science Museum, Simon Chaplin of the Hunterian Museum at the Royal College of Surgeons of England, Mark Nesbitt at the Royal Botanic Gardens, Kew and a host of staff at the National Maritime Museum, Greenwich, deserve particular mention.

A big thanks is also due to Alun Anderson for the original idea, to Jeremy Webb for his unwavering enthusiasm for these stories, and to *New Scientist*'s team of subeditors and designers both present and past. At Profile Books, Andrew Franklin, Paul Forty and Fiona Screen have gone to immense lengths to bring Farmer Buckley's trousers to a wider audience.

Last but not least, the editor would like to salute Mick Hamer for his stamina in reading several hundred 'Histories' over the past decade.

# Index